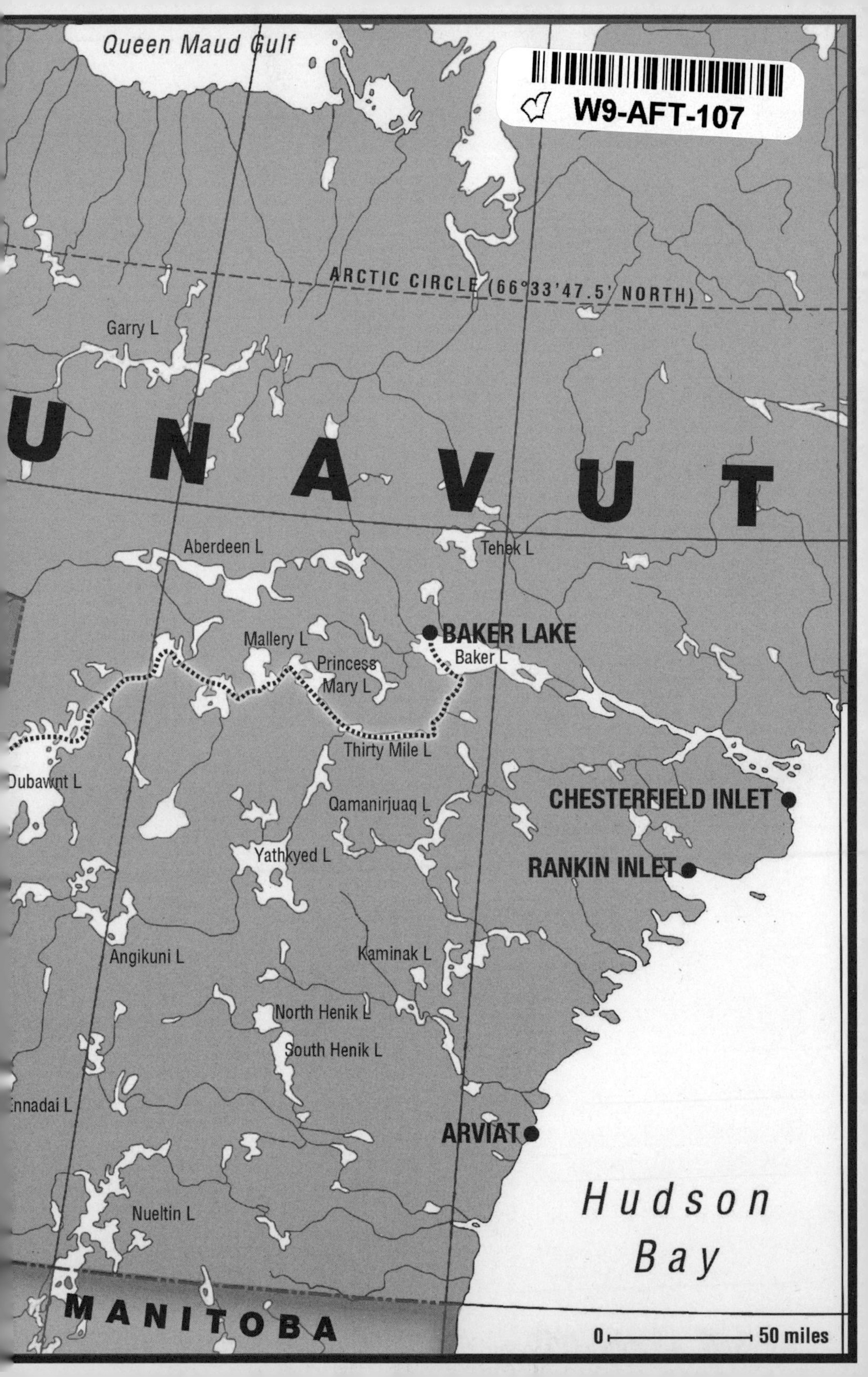
Queen Maud Gulf
ARCTIC CIRCLE (66°33'47.5' NORTH)
Garry L
UNAVUT
Aberdeen L
Tehek L
BAKER LAKE
Baker L
Mallery L
Princess
Mary L
Thirty Mile L
Dubawnt L
Qamanirjuaq L
CHESTERFIELD INLET
RANKIN INLET
Yathkyed L
Angikuni L
Kaminak L
North Henik L
South Henik L
Ennadai L
ARVIAT
Hudson
Bay
Nueltin L
MANITOBA
0
50 miles

Acclaim for Alex Messenger's

THE TWENTY-NINTH DAY

"Inviting and cinematic...The whole thing is told in exactly the same kind of upbeat...turbo-charged prose that filled, for instance, Aron Ralston's *Between a Rock and a Hard Place*...All that's left is for *Stranger Things'* Finn Wolfhard to step into the role of Alex and do his best with a CGI bear."

—*Open Letters Review*

"Almost fifteen years later, Alex Messenger revisits the canoeing expedition that in a matter of minutes turned into a life-changing ordeal. That distance lends honesty and verve to a narrative that steadily builds toward its gripping climax."

—DAVID ROBERTS, author of *Alone on the Ice and Four against the Arctic*

"I will never forget the first time I met Alex and his impressive telling of the bear attack. I was transfixed...And now with his whole story down in book form, I see his writing is even better than his storytelling! I was transfixed again."

—JIM BRANDENBURG, award-winning nature photographer, filmmaker, and bestselling author

"*The Twenty-Ninth Day*...well done! Alex Messenger is quite a good writer. It felt like I canoed the Barrens and long for a hot bath to soak my wounds."

—LONNIE DUPRE, polar explorer and author of *Alone at the Top: Climbing Denali in the Dead of Winter*

- A *Wall Street Journal* bestseller
- Finalist for the Minnesota Book Award
- A Midwest Indie bestseller
- An *Outside* magazine pick of best winter books
- *Paddling Magazine* recommended summer read
- *St. Paul Pioneer Press* selection of fall books from Minnesota authors
- *Duluth News Tribune* pick of "Outdoor Books to Read When You Can't Go Far Outdoors"

"A riveting read...The harrowing encounter with a grizzly bear and subsequent survival decisions take us straight to the heart of both life and adventure."

—HEATHER "ANISH" ANDERSON, author of *Thirst: 2600 Miles to Home*

"We were completely swept up in this story of wilderness adventure and survival."

—AMY AND DAVE FREEMAN, authors, educators, and 2014 National Geographic Adventurers of the Year

"One of the most harrowing first-person accounts of wilderness survival I've ever come across."

—CATHERINE WATSON, award-winning writer and writing coach

THE TWENTY-NINTH DAY

SURVIVING A GRIZZLY ATTACK IN THE CANADIAN TUNDRA

THE TWENTY-NINTH DAY

Alex Messenger

Published in 2019 by Blackstone Publishing
Cover and book design by Alenka Vdovič Linaschke

Printed in the United States of America

First edition: 2019
ISBN 978-1-9825-8333-0
Biography & Autobiography / Survival

3 5 7 9 10 8 6 4

CIP data for this book is available
from the Library of Congress

Blackstone Publishing
31 Mistletoe Rd.
Ashland, OR 97520

www.BlackstonePublishing.com

To my family, for showing me the world, and to my wife,
Lacey, for helping me see it all over again.

Bears are made of the same dust as we, and they breathe the same winds and drink of the same waters. A bear's days are warmed by the same sun, his dwellings are overdomed by the same blue sky, and his life turns and ebbs with heart pulsing like ours. He was poured from the same first fountain. And whether he at last goes to our stingy Heaven or not, he has terrestrial immortality. His life, not long, not short, knows no beginning, no ending. To him life unstinted, unplanned, is above the accidents of time, and his years, markless and boundless, equal eternity.

—John Muir

CONTENTS

PROLOGUE 1

PART I

1 Entering the Taiga: Days 1–2 7
2 Water: Days 3–8 17
3 Ice: Days 9–13 31
4 Dubawnt Lake: Days 14–16 43
5 The Final Push on Dubawnt Lake: Days 17–18 49
6 Dubawnt Canyon: Days 19–20 59
7 Ebb and Flow: Days 21–23 69
8 Leaving the Dubawnt River / Trailblazing: Days 24–25 79
9 The Kunwak: Days 26–27 87
10 Fate: Day 28 95

PART II

11 A Layover Day: Day 29, 10:00 105
12 A Chance Encounter: Day 29, 19:20 111
13 Awakening: Day 29, 19:31 123
14 Airway, Breathing, Circulation: Day 29, 19:41 131
15 What Now? Day 29, 19:45 141
16 The Evacuation Algorithm: Day 29, 19:50 149
17 The Night: Day 29, 23:00 167

PART III

18 Learning to Walk: Day 30 ... 175
19 Loss: Days 31–32 ... 191
20 Fear: Day 33 ... 209
21 Surgery: Day 34 ... 221
22 Evac: Day 35 ... 239
23 Premonition ... 253
EPILOGUE ... 257
ACKNOWLEDGMENTS ... 263
GLOSSARY ... 267
AUTHOR'S NOTE ... 273

PROLOGUE

The cool wind blowing across the subarctic tundra rustled the thin nylon of my tent, the crisp air refreshing and relaxing me. Sunlight shone through a thin haze of high cirrus clouds, gently warming the tent, adding to my sense of well-being. I was comfortable, cozily tucked away in my sleeping bag, and soon fell into a deep slumber. It was midafternoon on the twenty-ninth day of a six-hundred-mile canoe journey that I had embarked on with five other paddlers.

From the depth of a dream, I jolted awake, sitting upright and gasping, my mind lurching toward consciousness. My chest heaved, lungs filling as if I had come up from under a wave. My dream was already forgotten. Alarmed, I looked around the tent. In a place with no sense of time, I was overcome with the uneasy feeling that I was late for something.

Pushed by that unsettling sense of urgency, I threw on my clothes and shoes and left the tent, crossing the hill in long strides. I began climbing the steep, rocky ridge above our campsite. Still feeling late and shaking off the lingering fog of my nap, I was soon breathing heavily from the hard ascent. It took several minutes to gain the hundred vertical feet of the ridge. At the top, I found a lunar landscape of gently rolling granite domes scattered with smaller stones. I was four hundred feet above the wide blue expanse of Princess Mary Lake and could see thirty miles in all directions. The treeless vista was startling in its grandeur, the scale difficult to comprehend.

Barely able to take it all in, I turned toward another slightly higher point in the distance, topped with a rock cairn—an *inukshuk*—and resumed my hike. My Nikon, in its waterproof case, hung heavily in my hand. I decided I would leave it closed until I reached my destination. I walked past small round boulders, deposited along the ridge by lumbering glaciers eons ago. The large crescent moon of the island bowed in front of me, pointing north toward the Arctic Circle. I walked, watching the ground beneath me, the sparkles of sunlight in the granite and the green-gray blooms of lichen.

The feeling of tardiness slowly ebbed, and I began to calm down, my mind wandering. I thought about my coming senior year of high school and the summer reading book in my pocket, *The Liars' Club.* Quickly casting that thought aside, I thought of my camera in the case and imagined what I wanted to shoot first. I passed more lichen and granite, more scrub grass. Taking out the camera or reading the book seemed like too much effort right then, and the thought of sitting near the inukshuk seemed perfect.

Halfway to the cairn, I was still studying the ground, walking up one of the gentle granite domes, when something flashed in the upper periphery of my vision. My head snapped up. Thirty feet in front of me, at the top of the dome, an image materialized. Brown fur. My core tightened and my pulse doubled. All the muscles from my shoulders to my legs tensed at once. I thought of the ornery musk oxen we had seen earlier. We had been warned they were quite dangerous. As my brain decoded the firing synapses and visual signals, I made a far more horrifying realization. This was no musk ox—it was much worse. At this instant, the creature's gaze met mine. I was staring into the sharp black eyes of a grizzly bear.

PART I

> There is magic in the feel of a paddle and the movement of a canoe, a magic compounded of distance, adventure, solitude, and peace. The way of a canoe is the way of the wilderness and of a freedom almost forgotten. It is an antidote to insecurity, the open door to waterways of ages past and a way of life with profound and abiding satisfactions. When a man is part of his canoe, he is part of all that canoes have ever known.
>
> —**Sigurd F. Olson,** *The Singing Wilderness*

CHAPTER ONE

Entering the Taiga: Days 1–2

The plane vibrated with the cacophonous drone of idling engines. We had over an hour of flying ahead of us, and then forty-two days of paddling. The pilots, now directing us to climb aboard from their perch on the float, had filled the Twin Otter turboprop with our gear, from the forward bulkhead to the tight heap of large packs at the rear. Our three Old Town Tripper canoes filled the entire right side of the plane. The only open spaces left were the cockpit, the narrow row of six canvas seats on the port side, and the space above the poorly sealed spare fuel barrel stowed immediately behind the copilot's chair. The drum wafted fumes of jet fuel, which we tried in vain to ignore.

Dan was clambering over the seats in front of me, his six-and-a-half-foot frame scrunched by the small fuselage and the wall of canoes. The seats were austere—old canvas slung over aluminum frames, positioned in the small space like hurdles in a tunnel. We had to step awkwardly over each one until we found our spots. At an even six feet, six inches shorter than Dan, I was also having trouble traversing the seats. Behind me was Jean in his bright-red rain jacket, eyes hidden behind sunglasses, jaw tight as if to hold in the same nervous excitement I felt clenched under my ribs. Behind him was Auggie, the least bundled up of us, in just a few layers topped with a thin fleece. He was ready to go, calm. Behind Auggie was Mike, in bright yellow, his recently shaved head up, eyes open. Bringing up the rear was Darin. The littlest of us, he took the smallest seat at the rear bulkhead, behind a cache of gear and beside the large door at the back of the plane. His hair rolled out in dark curls from the sides of his black fleece cap, and he pressed his lips together as he located and clicked the buckle to his seat belt. I found mine and then looked out the hazy porthole, past the blur of propeller blades spinning loudly on the wing.

Standing on the dock was a group of five girls, the Femmes, who would be making a counterpart trip to ours. Their canoes and gear lay in a heap onshore, just as ours had. After setting us down in the emptiness of the Canadian taiga, the plane would come back for them. They, too, would land somewhere deep in the wilderness and paddle great rivers and lakes. Our routes would run parallel, hundreds of miles apart, until several weeks from now, when our paths would finally intersect. We wouldn't see them there, however; their route put them half a week and a hundred miles ahead of us by then. We wouldn't see them again until we met in the little town at the end of our routes, 550 miles almost due north from where we sat.

"See you at Baker Lake!" we all had said, and before I knew it, I was looking out at them from my seat in the plane. I molded my earplugs and let them expand in my ears, muffling the rumble of the idling engines. Out the porthole I watched as the smiling group lined up shoulder to shoulder to start the Camp Menogyn traditional line-dance send-off. The Femmes kicked together, left and right, singing "Happy Trails." We couldn't hear them over the roar of the plane, but we knew the words.

The engines throttled up, and the plane shook. Prop wash and lake water buffeted the farewell chorus. Grabbing at hats and sunglasses before they could fly away, the Femmes finished the song and disappeared from view.

A minute later, we were in the air above Lynn Lake, skimming the trees on the far shore. Soon, we were going nearly two hundred miles per hour, a thousand feet above northern Manitoba and the endless scrub forest of the Canadian taiga. I looked at the shimmering amoeba-shaped blobs below—nothing but lakes and unturned land in all directions, as far as I could see. We flew for over an hour, the hum of the engine, the rattle of the fuselage, and the acrid smell of jet fuel feeding a growing headache.

When we arrived at our destination in the middle of nowhere, we circled above Wholdaia Lake. Our pilots made a scouting pass, looped around again, and dropped in for a smooth landing on the first lake of our trip. Using the propellers, they maneuvered their ungainly boat to shore, backing the pontoons onto the rocks and beaching the plane with a violent shudder.

The six of us wriggled out the rear door and took our first breath of fresh subarctic air before unloading the boats. It had felt odd to disassemble our canoes before the flight, stripping their yokes and thwarts so they would nest and all three would fit. Now the pilots helped us pull the enormous canoes out of the tiny plane, tightrope-walking along the pontoons. I wanted to get the boats back together before the pilots left, in case we ran into trouble. These canoes were our lifelines. Without them, we'd be screwed: hikers with canoe gear.

The blackflies found us immediately. They came silently, tiny malicious dots that attacked exposed skin at the wrist, waist, and hairline, burrowing their heads and leaving hot welts and weeping trickles of blood on our skin.

Soon, all our gear was out of the plane, and the pilots were preparing for their flight back to Lynn Lake from the middle of an almost empty map. By the end of the day they would probably be back at some camp, some town, maybe even in their own beds. We were in a strange, foreign place and would be for another month and a half, but to them we were just a charter penciled on their calendar. While they might hardly remember us in a week, this was the beginning of something life-changing for my fellow paddlers and me—this was Hommes du Nord, forty-two days of canoeing Canada's wild rivers and lakes.

I was seventeen.

Six months earlier, I was kneeling in the snow at Caribou Rock overlook, above West Bearskin Lake in Minnesota, a few miles south of the Boundary Waters Canoe Area Wilderness and the Canadian border. It was just after Christmas 2004. Beside me were Mike, a new friend I'd met just days earlier, and several other campers and guides from Camp Menogyn's winter camp. Our breath puffed clouds in the still winter air. The pale-gray sky dropped thick white snowflakes that floated like goose down, nearly weightless.

I had to decide.

"You should come," Mike had said. Going on Hommes du Nord meant spending forty-two days traveling through northern Canada—forty-two days on trail, forty-two days of white-water canoeing, portaging, sleeping on a thin mat in a thin tent, forty-two days of dried food, forty-two days of adventure and fresh air.

I had spent the past three summers working my way through Menogyn's trip progression, going on ever-longer and more intense paddling adventures until finally being invited back for this last big trip, the longest and most remote that the camp offered. I had the requisite experience, though not the money it would cost. I thought for a long while, watching the silent snow drift down.

Kneeling on the cushion of sphagnum moss, I finished reattaching the thwart to our canoe and looked up at our bush pilots. These might well be the last other humans we would see for six hundred miles. My stomach knotted up at the thought of our impending solitude. We were getting the last of the canoes together; it looked as if they would work after all. We double-checked the plane to make sure we hadn't forgotten anything, but all that remained aboard were the seats and the acrid-smelling fuel drum. We were good to go, and the pilots were ready to leave us behind. They latched the Twin Otter's rear door and walked along the narrow pontoon before mounting the ladder to the cockpit.

"Well," the pilot said, "enjoy the flies. I think you're crazy, but have a safe trip!"

He and his copilot ducked inside, whirred the propellers, and screeched the pontoons off from shore, slowly taxiing out onto Wholdaia before pushing the throttle and accelerating. The plane lifted into the air, the growl of the engine wafting over the water. Gaining altitude, they slowly arced back around toward our group. As they gained speed, they decreased their altitude to just ten feet above the water and headed straight at us. The nose of the plane got bigger and louder until it looked as if it might slam into us. I felt that I could jump and touch one of the floats. We ducked and yelled, waving and cursing. At the last moment, it pulled up, the air vibrating with the engines' roar as they buzzed the scrubby trees around us. I'm sure the pilots were snickering in the cockpit as we wiped lichen and moss off our clothes and out of our hair. Our trip was barely underway, and the excitement had already begun.

As the sound of the plane disappeared to the south, we walked up a mossy berm to a spacious clearing away from the shore. In a huddle, we talked about the trip and how we were feeling now that we were actually on trail. With the plane's departure, my stomach had begun churning with nervousness. I'd been preparing for the trip for six months, knowing I would be alone with a small group deep in the northern Canadian wilderness. Still, until the plane disappeared from sight, it hadn't hit me what this really meant. Suddenly, we were alone. We had a satellite phone, but if the need arose, help would be hours away at best, and at worst, completely out of reach.

We could rely only on ourselves and our group, but most of all, we were reliant on Dan. I looked at him now as he led our discussion. He was ten years my senior and had been guiding canoe trips forever. I had full confidence in him. His skill, expertise, and leadership would get us down the Dubawnt River, to the Kunwak and Kazan rivers, to Baker Lake, and eventually home again. It was at first disorienting to be suddenly in such deep wilderness, so completely removed from our usual ways of life, away from the safety nets, everyday comforts, family, and friends. We all were seasoned campers, but this was different. The Canadian subarctic is a kind of wild where you can expect not to see anyone else for weeks, and the land seems to stretch on to infinity. This would be the longest trip yet for each of us. We were already deep into the taiga, and the only way out was six hundred miles of paddling and portaging north to the end of our trip at the tiny, secluded town of Baker Lake.

We huddled there for a while, the taiga silent but for the thick clouds of blackflies and mosquitoes buzzing around us. I felt odd, unsure of where I was or where I was going. I imagined the sweeping tundra as it extended north and east to Baker Lake, six hundred serpentine miles of rivers, lakes, and portages. The whole expanse between where we stood now and that tiny dot of a town was all but blank in my mind—just hazy imagination and the obscure translation of land and water onto paper. That blankness made me uneasy, but I managed to push the feeling aside as we packed up and prepared to hit the water.

Despite the nervousness that I was sure we all felt, that first day was beautiful, with long hours of sun, water, and paddling. We made excellent time and then napped lazily in the boats. By the time we camped, we were far along the Dubawnt River, already on our third topo map and two days ahead of schedule.

We set up our two tiny MSR Prophet expedition tents but decided not to set up the colossal bug tent, which we'd brought for cooking and relaxing. Once camp was mostly prepared, Darin and I fished for a while, but only caught northern pike. We weren't fond of handling or filleting the slimy, toothy devils, and we had enough food in our three seventy-five-pound barrels to last the whole trip, so we threw the pike back.

When we got back to camp, the mother of all mosquito swarms joined us for macaroni. I swatted at them and walked backward and forward to avoid them. Still, my macaroni was dotted with lifeless black lumps of insect protein. In the far north, food is a finite resource. I tried to pretend the dark little dots were black pepper, and kept eating. Next time, pain though it was to set up, we would pitch the bug tent.

Our second day on trail, we paddled from morning until night on the Dubawnt. As we made our way downstream, arctic terns flew overhead and buzzed the water, darting back and forth. They were astonishingly agile and fast, their black-and-white feathers appearing and vanishing so suddenly they looked silver.

But for the sharp whoosh of tern wings, the occasional breeze, and the murmur of running water, silence held us in its thrall. This was not the stuffy silence that happens in a space closed off from the outside world. Silence in the wilderness is the raw sound of vastness. It made me feel small, inconsequential. I imagined looking into the mouth of a cave. Deep, dark, and mysterious, it might go on forever, or it might collapse at any second. I had felt this before in northern Minnesota's boreal forests, overcome by the sound of an entire ecosystem, a deep hum almost imperceptible to the human ear. At first, it had been overwhelming, disorienting even—but I had slowly become comfortable with the feeling. In this grand space of the Northwest Territories, it was magnified. As I'd grown accustomed to it, it had always surprised me how natural it felt—this sense of a place infinitely older than I, bigger than I. It was rejuvenating. Maybe that was why I signed up for this trip.

We paddled downstream, over lakes, around rocks, and through small rapids until we finally reached the lake that we would call home for the night. There, the wind picked up. When lightning exploded from the sky, we turned to shore. Within minutes, we were struggling to set up our tents as the trees around us bent in the stiff wind, sand stinging our faces like blowing sleet. The shelters whipped like huge kites as we tried to grip their corners and stake them down. I pictured one ripping from my hands, flying out into the lake before it sank in the water or tumbled on across the emptiness. The thought made me shiver, and I drove the stake a little deeper.

With the tents pitched, we threw our sleeping bags and clothes inside, tucked the canoes safely in from shore, and stowed our mostly empty packs under them. Confident now that nothing could blow away after our bombproofing, we took cover in the tents. It was July 4, 2005, and as we sat listening to the rush of wind and rain and the roll of thunder, I thought of friends and family at home, experiencing fireworks of a different sort. As the tent swayed around us, I thought of thick green grass on rolling hills, the soft pile of a blanket spread on the ground, and the collective *oohs* and *aahs* from faces illuminated by every color of fire.

Our tent flashed pale yellow, followed by a loud *crack!* Not a second between the light and thunder. The bolt must have struck the far shore. My daydream faded quickly, and I pulled my feet even higher up on my sleeping pad to lessen my conductivity with the ground. I stayed like that, waiting for the storm to pass.

The wind gradually calmed, and then the rain stopped. I climbed out and found evening sun emerging from the retreating storm clouds. Behind our site, a long, winding ridge ran parallel to shore. I hadn't noticed it before—a smooth slope of gravel rising to a narrow ridge. It looked out of place—manufactured, even. It reminded me of a bike trail I'd ridden countless times as a kid—a decommissioned railroad grade that had been converted into a multiuse trail that wound above swamp and through thick woods, west from Minneapolis to the small town of Excelsior. It was a four-mile ride on a hot summer day to get the best frozen custard in the state.

Like the bike trail, this ridge behind our site was the same height as far as I could see, the same steep slope rising from shore. It was an esker, a remnant from an ancient glacial river, and the first I'd seen. When the ice melted, the sediment from the bottom of one of those blue rivers had come to rest on the ground to form this huge snakelike mound. These raised veins and arteries left over from the Ice Age were all over the northern Canadian Shield. Our maps showed eskers with small parallel hash marks like rows of wooden railroad ties, as if depicting a long, twisting track that was never finished, the rails never spiked down. Northern Minnesota is covered with evidence of the glaciers, too: bedrock carved into smooth undulations, house-size boulders abandoned by the receding ice, and thousands of lakes running parallel on the map, as if gouged by a giant's claws. But the timeless processes that left those indelible marks on the landscape seemed somehow abstract, a mere intellectual correlation. This esker, though, fascinated me. Water flowing through a channel that itself is made of ice seems a temporary thing, a powerful river living a fragile existence, one that was ended by a few degrees' rise in temperature. Yet here it was, the wandering line of a river, complete with curves and eddies, the stones and boulders once caught in its flow now a permanent part of the landscape.

Climbing the esker, I could see the broad, glimmering expanse of the lake, a pale pink sun, and the dark steel gray of the disappearing storm. It was a beautiful evening. Since starting on Wholdaia Lake, we had been paddling the seemingly endless expanse of bushes, spruce, and fir that forms the Northern Canadian Shield taiga. This lake, though, with its eskers and sandy beaches, felt different from others I had seen. The ground felt worn in a way it hadn't before. The sky seemed bigger, the water colder. I felt as if we had paddled back in time.

> I have never yet seen a river that I could not love. Moving water…has a fascinating vitality. It has power and grace and associations. It has a thousand colors and a thousand shapes, yet it follows laws so definite that the tiniest streamlet is an exact replica of a great river.
>
> —**Roderick Haig-Brown,** *A River Never Sleeps*

CHAPTER TWO

Water: Days 3–8

Over the next few days, we eased into the groove of the river, getting comfortable with the way the water moved around rocks, with what we could see and what we couldn't, with what we were capable of as individual paddlers and what we could achieve with our paddling partners. Each day we changed partners, so every set of rapids was different, both because of its unique features and formations and because the dynamics of the boat changed from day to day.

We were paddling white water conservatively, meaning that our objective was to take the cleanest line downstream. We weren't aiming for thrills or big waves, trying to surf the canoes, or anything of the sort. Our goal was to make it down to flat water again with the least amount of drama. Despite our best efforts—and not to anyone's disappointment—this still left plenty of room for excitement.

Many trips on the Dubawnt were outfitted with bow or full-canoe skirts—waterproof fabric tarps custom-fitted to the boat, which would prevent water from coming in over the sides. We had no such precaution; they were intentionally left behind in the hope that it would encourage us to paddle more cautiously. The fear was that having the skirt would embolden us, that we would paddle harder rapids, hit bigger waves, and take more risks. The fact was, without the protection, we were more prone to mishaps arising from small mistakes. A bad angle on a ferry, a wave that proved a bit bigger than it looked from upstream, or a chute that dropped us down with more energy than we'd thought—any one of these could dump buckets of water into the boat. Enough water over the gunwales, and the canoe would start to get clumsy, not responding to the paddlers' actions. More water still, and it would submarine, sinking lower and lower until all that remained out of the water were the paddlers themselves—from the waist up. At that point, there is not much left to do but hope for shallow water or an assist from one of the other boats.

I don't think there was anyone who didn't like canoeing white water. The feel of the boat being pulled downstream, the power of the river, and the sensation of walking a tightrope between clean-and-safe and the chaos of a swamped canoe was addicting.

The far-off rumble of water over stone was like a drug. It filled me with nervous excitement. Goose bumps rose on my arms as I subconsciously held my breath, turning my head to listen, as if by hearing the rapids I might read their waves, find our way through before even seeing them.

White water wasn't new to any of us. The year before, we each had spent a month paddling rivers in the southern Canadian wilderness on trips called Norwesters. Before beginning this trip, we had trained again on the rapids of Minnesota's St. Louis River, reviewing our basic paddle strokes, the draw and pry to pull or push the canoe sideways, and the J and C strokes for controlling it from the stern. After practice-spinning our boats in a bay, we had moved on to more specific techniques for white water, ways of moving the canoe that were as choreographed as a dance and described by terms just as foreign and arcane. Dan would call out strokes. "Cross draw!" and we'd reach across with our paddles, and the canoe would turn in the water. "Draw!" and we'd switch back, spinning the boat the other way. We launched the canoes down short sections of river and whirled them around into the calm eddies with a stroke called the *dufek*, an aggressive bow-rudder that made the boat turn on a dime. Where the risk was too great to paddle, we practiced controlling the boats externally with ropes, from shore or shallow water—a technique called *lining*. After our review, we paddled the St. Louis from Interstate 35 to the Thomson Reservoir, 4.1 miles of class I–III rapids with names like First Hole, Hidden Hole, Rescue Rapids, and Last Chance Rapids. Our training made us confident and reminded us of the thrill of the paddle. But all of that had been without packs, tents, and barrels of food, in warm weather, in a place where cell phones had service and scoops of ice cream were just a short drive away. The Northwest Territories had none of that fallback, that safety net. And the Dubawnt River was much bigger and more intense than the St. Louis.

Early on day four, I was reminded how quickly things can go wrong in rapids. Halfway through a set, Jean used a draw instead of a cross draw, and the bow of the canoe swung suddenly in the wrong direction. A second later, we slammed broadside against a rock and beached high-centered on top of it. We were stuck bobbing with the waves like a turtle on its shell. To free us, I had to step out onto the rock and push off. The boat lurched downstream, and just when I'd managed to scramble back on board, we slammed into another rock and I had to push off again. Backward now, we twisted around, doing evasive draws and pries to avoid the rocks. Finally, we slid backward into the calm eddy at the bottom, where we found the other guys waiting behind a boulder, paddles across their laps, eyebrows raised when they saw us.

"Don't ask," I said.

We were lucky. A misstep like that on a big set, and we could easily have swamped or worse. If the features had been bigger and the current stronger, our boat could have wrapped like tin foil around the rock. We were lucky, and we were humbled.

Although paddling white water is inherently dangerous, we did all we could to manage the risks. We always wore personal flotation devices, and we'd trained in throwing ropes to rescue a teammate floating downstream. We had been warned about the dangers of foot entrapment, a terrifying scenario in which the victim's foot catches under a rock or other obstruction and the current pulls the rest of the body under downstream. To avoid this, if we ended up in the river, we were to turn and face downstream, float our feet up in front of us, above butt level. It made for a bruised backside but kept our feet safe from snags. Strainers were another danger. Trees are the most common type of strainer, their branches or roots acting like a sieve in the water, trapping debris and paddlers alike. The force of the current on a strainer can easily trap a person, and we knew to avoid them at all costs. We also had to watch out for recyclers, tumbling waves that roll over and over again, seemingly without end. Objects entering a recycler tend to do the same. Paddlers, even boats, can get caught in one, tumbling like a wet load of laundry. We'd been told that if we found ourselves in a recycler, the best way to escape was counterintuitive: to go spread-eagled in the hope that an appendage would catch a nontumbling patch of water and help pull you out of the maelstrom. If that didn't work, we were supposed to ball up and sink, or swim out, up, down, to the side—anywhere away from the tumbling torrent. Still, the most important way for us to manage these and other dangers was to avoid them altogether through scouting, communication, training, and practice in handling the boat so that it stayed upright with us in it.

But the river wasn't the only place we were exposed.

On day five, as we made our way downstream, we monitored a storm system that had been boiling up in the distance. It had looked as if it would miss us, but as we started down from an eddy, the sky began to grow dark. The storm overtook and quickly surrounded us. The river turned black, and the rain came in sheets, followed by a beat of no rain before the next onslaught. Suddenly, lightning cracked loud and close. As we turned frantically toward shore, a pair of otters appeared in front of our canoe, bobbing on their backs in the dark river and looking quizzically at us. They seemed to smile, as if persuading us to stay. We ignored them and dug our paddles in hard, getting to shore as lightning and thunder came only a few seconds apart.

Scrambling onto land, we heaved the boats, still fully loaded, as far as we could onto the rocks, then fanned out across a hollow that was filled with stunted sumacs and squatted on our haunches. Behind the hollow, the graceful rise of a short esker formed a protective curving berm. Flashes and explosions of thunder came at once and from all sides. Crouched in the lightning position we'd been trained to assume, with enough distance between us to avoid multiple casualties from the same strike, I flinched as dazzling bolts forked across the sky and made the air tremble. Some of our group watched while others chose to lower their heads into their hoods. The relentless downpour continued its assault while we each endured the barrage on our own.

Lightning was one of my biggest fears on trail. When I was still in diapers, our house had nearly gone up in flames after a lightning strike, the fire extinguished only after repeated dousings with water from my emptied diaper pail. My great-grandfather had been struck by lightning in Cleveland, Ohio, and the story of his white long johns scorched in the pattern of his blood vessels passed from one generation to the next. My father's advisor and mentor in graduate school had been struck by lightning atop El Castillo in Chichén Itzá, and died when my dad's CPR proved unsuccessful. It is a powerful, indiscriminate force. On the water, the lightning made me skittish, hurried, almost frenzied. Now on land, I tried to calm down—I told myself that I could do only so much and that the rest was out of my hands. It may seem like a passive attitude, but once we were in our lightning positions, there really wasn't much else to be done. We waited as the rain buffeted our crouching bodies and the sky lit with the sharp claws of electricity.

I thought of the good-luck charm I had, wrapped in a round Altoids tin in my dry bag. A gaunt old monk in Thailand had given it to me. I could still see his thin arm outstretched beyond his saffron robe, the tiny purple amulet looking heavy between his bulging knuckles. I remembered a relevant lesson about worry and the needless suffering that it causes. We worry about so many things in life, and in most cases, there is no reason for it. If something we're dealing with is worrisome, better to transfer that worry into action and confront or prepare for the circumstance. If it is out of our control entirely, then we can't influence the outcome anyway, no matter how much we let it consume us—so we shouldn't let it occupy our minds.

Still, having done everything I could to stay alive, I didn't feel that kind of peace. I double-checked that I wasn't near any natural lightning rods or perched on an exposed root bed, and I adjusted my squat so I was as small as I could be and not touching any more ground than necessary. With the world exploding around me, I tried to take a deep breath and let it out slowly. I had done all I could to protect myself.

Finally, almost as quickly as it had come, the tumult of the storm passed, and before long it was barely an echo in the distance. The sky brightened. We rose stiffly from our squats, our rain gear soaked and dripping. I felt relief and the shaking uneasiness of ebbing adrenaline.

That night, from our camp on a tall gravel hill, I could still see the storm on one side of the vista while hazy beams of sunshine filtered through clouds on the other. In every direction were bodies of water large and small, and endless rolling hills with black-and-pink bedrock peeking through the green velvet that carpeted their slopes. Dwarfed pine trees made everything seem even larger than life. It felt like looking across an endless diorama where everything was perfectly, carefully placed to create scenes within scenes. The rolling hills revealed and obscured distant dark waters while the shafts of pale-yellow sun shifted slowly from one spot to another.

Storms, we'd been told, were rare in the taiga.

The next day, we kept a wary eye on the sky. It felt at once close and expansive, like peering up through a fish-eye lens. We could still see the weather we'd had the day before, slowly plodding east while a new storm system brewed to the west. Over the course of the morning, those dark clouds drifted past. Then shadowy veins splintered out across them like cracks through a frozen lake. A dark curtain of rain fell from the new opening, and rainbows glowed around it. From a welcome distance this time, we watched as spectacular lightning bolts flashed in silence across the sky.

Later in the day, we approached a set of rapids and pulled off the river to scout ahead. We walked a quarter mile alongshore and climbed a granite glacial erratic. The huge boulder had been dropped here eons ago by a glacier. It stood some fifteen feet tall and was covered with leafy brown rock tripe lichen. From the top, we could see down another quarter mile of turbulent, boiling features to where the river made a sharp ninety-degree turn at a boulder, as if the rock were the fulcrum on which the whole river hinged. This set would be a challenge.

Back in the boat, Darin and I knelt on the pads and started downstream. Halfway down the set, we shot for an eddy and missed. Instead of pausing to regroup in a calm pool, we were suddenly paddling hard against the current, with no eddy in reach, again going down a set backward. But this was a much bigger set than the last. Running the next section backward through hard rock dodges, steep drops, and the sharp curve around the fulcrum boulder would most certainly cause a swamping. We looked around frantically and found another tiny eddy downriver. We ferried over and slid into it. The moment we crossed the line into flat water, everything slipped into slow motion, and we drifted calmly as the river whipped around us. We steadied our breath as we surveyed the river and the course of our next run downstream.

After a minute of bobbing in the eddy, we paddled hard and fast to rocket out of it. Crossing into the current, I dug my paddle in and we spun around, leaning the boat hard as we entered the river. Instantly, we were in the thick of the white water again, dodging spray. We were coming to the last and trickiest part, which included narrow Vs, big standing waves, and the tight ninety-degree turn with the fulcrum boulder on one side and a roaring, tumbling recycler on the other. It was a narrow chute with hazards on either side and more boulders just beyond. Clearing the last boulder before the fulcrum, we pulled hard, working in concert to turn the canoe. The boat almost spun, the turn was so quick. The current pulled us toward the deep chute. Darin and I leaned into the turn as strong lateral forces nearly whipped us and our gear out of the boat and into the river. In an instant, we pulled our arms in to clear the boulder and dug in again to keep paddling, keep bracing, and keep steering. Another instant later, we had splashed down the chute into calmer waters below. My blood pumped, whumping in my ears with each pulse. We were the last boat; all three had made it free and clear. We whooped and congratulated each other on the clean run. What a rush. It was one of the fastest runs I had ever done. Our open band of safe water couldn't have been more than five feet wide, barely wider than the boat. On either side were features that could have destroyed us. It was like running along a narrow mountain ridge—don't go left, don't go right. *Whoosh,* right through the middle.

The next day, when we reached Barlow Lake, Mike clipped a classic white-and-red-whorled Dardevle spoon to the end of his line and dropped it in the water. It splashed heavily and dropped deep. It wasn't long before the rod bent and Mike yelled that he had a fish. With such heavy line, he didn't need to use finesse, and soon he had hauled in our first lake trout. Its back was the color of dark steel, its sides covered in hundreds of tiny spots, its belly almost white. It was a big fish, built stout, with a pronounced upper lip and underbiting jaw that gave it a scowl. It was probably fifteen pounds and produced two beautiful pink fillets, each more than an inch thick.

We made camp a few miles from the end of the lake and carefully pitched the tents, fly first, in another rainstorm. We added the trout to a flavorful dinner of couscous and falafel. As we cooked, a rich red sunset burst through the rain, and full double rainbows arced across the sky. They shone so brightly that everything as far as I could see was bathed in a deep saffron glow.

It was still raining when we turned in for the "night." The sunset faded into twilight. Soon, dawn would come. There was no darkness—only degrees of shadow. It enveloped the tundra, our boggy campsite, and our tents. Still, for the first time on the trip, I had to write in my journal using my headlamp. The shadow was growing, and with each passing day, the dark of night was deepening.

On day eight, our paddle-partner cycle had circled back, and Dan and I were once again sharing a canoe. We paddled the rest of Barlow Lake in balmy sunshine. Auggie, in the boat next to ours, held the map close to his face to read the small contour lines on the lakes. Between Barlow and the next lake, several miles downstream, the river showed a steady drop. "Four-point-two feet per mile," Auggie said, his rarely seen braces glinting as he grimaced under the bright sun. A grade like that meant relatively gentle water with some mild rapids. We slipped off Barlow Lake, back onto the river.

Our calculations were wrong.

Instead of dropping over the course of a few miles, most of that elevation change happened within a half mile. We scouted, then ran the top part of the rapids before eddying out partway through to scout again. Even from shore, the rapids looked technical. Everyone was going to watch Dan and me run it first. I was nervous. We got back into the boat, sped up, and peeled out of our eddy, aiming for a patch of calm next to a tumbling recycler. We overdid it but quickly corrected. Barely past the recycler, we had to make a sharp right turn in fast current, then a sharp left, then hit another downstream V and rock-dodge from there. It was fast, but our boat slid gracefully back and forth through the slalom, all the while bounding violently with the waves. It felt as if we were cheating, slipping through an obstacle that was put there to keep us back. But we found a secret way through and made a clean run. It was exhilarating.

A few hours later, we reached Carey Lake, and it got suddenly colder. I took my waterproof GPS, which had a built-in thermometer, and lowered it into the lake on a steel fishing leader. After a few minutes, the screen read thirty-two degrees Fahrenheit.

We paddled quite a distance across the lake, into a slight headwind, before taking a break. Abruptly, the wind swung 180 degrees and pushed hard at our stern. Cold wind blew at our backs while dark clouds approached from the front, as if a frigid undertow were pulling us into the storm. We doffed caps and tied down anything loose. Shortly, we were hurtling down the lake, surfing three-foot cresting waves. We hugged close to the safety of shore, but soon we were riding the waves over a reef of pink, red, and gray stones that broke the surface all around us, somehow missing the boat.

"We haven't hit any rocks yet!" Dan yelled above the wind.

A rock appeared in front of our boat. "Dan!" I yelled, as I pulled hard at a draw.

The canoe beached halfway up the rock, nearly throwing us onto our hands and knees with the sudden stop. The next wave lifted us off and pushed us back into open water. The other boats had started catching rocks, too, and we moved out just far enough from shore to miss them. It was precarious, though. We had to get off the water. We pointed our boats to an island across the bay and rode the waves onto its shore to make camp. Tension had been high thanks to the white water and the unexpected wind, and it remained high as we quietly moved gear and set up the tents.

We began work on dinner, gathering our food barrels and the wanigan, the rigid cooking-cupboard backpack that kept our spices and cooking supplies. We pulled bags of pasta, rice, and sauces from the barrels, and finally Darin found the sauce we'd premade for that night's dinner, an Asian-inspired peanut sauce we called *gado gado*. Now he rummaged for the bag of pasta shells that would form the base for this and another meal. When he found the bag, we made a disquieting discovery. It was a small bag. After this meal, we had only half the shells we needed for the next one. Something wasn't right. "Pull it all out and count," Dan said. We carefully divided all the bags and portions into piles on the ground. The pasta shells weren't the only problem. We were low on other staples, too. We stared quietly at the mound of food, wishing there were more.

We had expected our caloric intake, high as it was, to be less than our output. Somehow, though, we had fumbled when packing out in that frenzied hour at Menogyn, mismeasuring or miscalculating as we tossed meals together and poured pasta, powders, and dried ingredients into bags, mixed the rich trail mix we called *pemmican*, and made *matt food*, a protein-rich concoction of peanut butter, honey, powdered milk, and oats, which we ate with lunch. We had packed pantry style, with big bags of pastas and vegetables rather than individually wrapped meals. It was all divided into three distinct fifteen-day loops, each with the same set of meals and the same portions, packed into blue waterproof barrels. Each barrel carried seventy-five pounds of food, and yet we had missed a couple of scoops here and there. Instead of having six cups of noodles for a meal, we had maybe four to five cups. This wasn't a great shortfall, but given the number of calories we were consistently burning, it made a difference. Gradually, our bodies were eating away at our fat deposits. If the deficit grew too great, our bodies would begin to pull needed calories from our muscles. We were lucky the fish had started biting.

Disheartened by the prospect of short rations and exhausted from the day, most of us broke off to be alone. Mike, always active and doing what could be done, took a pot to pick wild cranberries. Darin turned inward, as he often did, and read one of his many books. Auggie tended the stove. The wind had died, and Jean and Dan tossed a Frisbee. I took a long walk around the island. It was nearly a perfect circle, the ground rising several feet above the water before it planed off at a low plateau. The ground rolled with ancient lichen-covered stones, a carpet of Labrador tea and cranberries growing between them. Other than the underbrush and a few stunted evergreens alongshore, vegetation was sparse. Without the wind, the only noise was the distant sound of the Frisbee thudding into the ground, and Jean's and Dan's voices wafting through the stillness. The island felt as deserted as a gray stump in a clear-cut. It made me uneasy. I had the eerie sense I wasn't alone.

On the far shore, four miles away, a tall hill climbed steadily up from the water. It was barren but for a stack of stones that rose like a statue from its peak. It was a cairn, distinct even from this distance. Human made. As hard as it was to imagine in this remote place, people had been here long before us. Cairns in the north are not built on the whim of tourists seeking balance on pebbled shores. Here, they were built for a purpose—for navigation, for funneling caribou for the hunt, and sometimes as a memorial. I looked at the stark silhouette in the distance. *Why are you here?*

Before long, Dan padded quietly through the low-growing Labrador tea bushes and found me near the pines. "Did you see *these*?" he asked, gesturing at the ground. A group of stones sat half-buried, covered in blooms of lichen. "They're in a circle," he said.

My eyes traced the arc of the stones, recognized they were placed at regular intervals, and saw the overgrown bald between them. I tried to imagine their cause: perhaps the cycle of freezing and warming, the push of ice and wind. But it didn't add up.

"What are they?" I asked.

"Tent rings."

I suddenly realized that I'd seen them all over the island. I thanked Dan for the insight, and he left me with my thoughts. *Tent rings.* A camp, not so different from ours, but from long ago and much larger. I remembered our food barrels. The people that made those rings had not had bags of pasta. They were living off the land, hunting caribou. We thought of ourselves as self-sufficient, which we were—an unsupported expedition with no food drops or resupplies—but our survival was predicated on the things in our boats—the gear and the food we had with us. Our contingencies should these be lost were few. Since these rings were built, many modern expeditions had paddled this lake before us, most without major incident, their details lost to history because everything went according to plan. One of the ways an expedition tests your mettle is that it's up to you and the group to solve any problems. If you wait for help to suddenly appear, your odds are not good. And if you pick up the satellite phone and call for help, it could still be too late. You have to stop the bleeding, splint the break, patch the tent, splice the cord, repair the stove, recover the boat, ration the meals, catch the fish, guard the maps, and protect the satellite phone. I shook the thought away. We were self-sufficient, just in a different way, with different parameters and equipment—a different purpose. I looked back down at the ring in front of me. *What brought these people here?* I wondered. *Did they build the cairn? Why did they leave?* I imagined a tan patchwork of caribou hides over bent wood. I thought about how I came to find myself on this very island. I had come for adventure, for the excitement and experience, but also, I was following the path still unfolding. I'd wanted to see what lay around the bend, where it all might lead. After my Norwester, I had decided to hang it up, accepting that I'd miss whatever view waited just over the next rise, and the next. Then Mike had convinced me it was worth seeing. Now here I was, looking at these ancient stone rings. I didn't know what it meant. I had more questions than when I'd started.

The evening deepened into a sunset that burst across the sky, bathing our little island and the taiga in a deep, saturated red. Sunsets in the north are not like those farther south. They last for hours as the sun dips slowly below the horizon only to rise back up again in a strange play of time and twilight.

Looping back toward our tents, I couldn't shake the uneasy feeling that I wasn't alone. The cranberries at my feet glowed in the light, the Labrador tea flowers turning from pale vanilla cream to deep ocher.

I nearly stepped on it. It was the same hue as the flowers, but a different shape. Feathers were spread wide as if to take off, but there was no body, only the wings. The ground was stained between them. Not long ago, it had been an arctic tern; the sharp angles of the wings were unmistakable. I wondered, what killed this bird? Then I remembered seeing scat and prints all over the island—evidence of animals much larger than this bird. And that uncanny feeling that we were visitors to someone's—now something's—home.

That was the first night we kept the bear mace in our tents.

Directly in front of me was the ice-covered bay of a great lake, and on the far side of this bay was something which at least relieved the somber monochrome of the muskeg colorings. It was a yellow sand esker, rising to a height of fifty or sixty feet and winding sinuously away into the distance like a gigantic snake.

—**Farley Mowat,** *Never Cry Wolf*

CHAPTER THREE

Ice: Days 9–13

Two days later, the river poured us out onto Nicholson Lake, a twelve-mile-long body of water with a narrow middle, and huge bays that swooped from either end, so that it looked like a treble clef on the map in my lap. After paddling across the first bay, we stopped at a small island for lunch, and Dan and I reviewed the chart. At a scale of 1:250,000, an inch on paper represented four miles, and the island we were on was barely a pinprick. It was so small, in fact, that at first we didn't even see it.

We had two full sets of maps for our entire route from Wholdaia Lake to Baker Lake—two thick packets in separate dry bags, stored in separate packs, always in separate canoes. These were arguably the most important of all our equipment. Without our maps, we would literally be lost. At the moment, as the crow flies, we were 240 miles from the nearest villages of Baker Lake to the northeast, and Stony Rapids to the southwest. We were 300 miles west of Hudson Bay, east of Great Slave Lake, and northeast of Lake Athabasca. Between us and those tiny hamlets of dirt roads and ATVs was nothing. Pure wilderness. I was 1,263 miles from home.

As we paddled across the wide mouth of a bay, a far-off thrum whispered over the wind. We paused, scanning the skies. Through the heat and mirage lines across the lake, the silhouette of a helicopter passed behind twin peaks of a small mountain. We slowed our paddling and watched as it swung steadily around the sky, rotors beating the air. This was not something we had expected to see. Float planes, sure, but no one sees helicopters this far out in the middle of nowhere.

The helicopter made a slow arc across the horizon. We had just decided to count the intrusion as a direct human encounter in our unofficial count of days without seeing anyone, when the helicopter changed course and swung back. To our surprise, it headed directly toward us, as if it had found what it was looking for. We looked across the water at one another, confused and concerned. I looked behind me at Darin. He tracked the aircraft, then looked to me, his cheekbones casting shadows over a frown, and shrugged as if to say he was as baffled as I was. Dan was worried. In the event that something dire happened with one of our families back home, this was one way we might find out.

The Bell four-seater got lower and louder until we could feel the rotor wash as it landed on a point only about a hundred feet from where we sat in our canoes. We paddled in to shore, and two men stepped out of the helicopter and came toward us, leaving the pilot at the controls as he slowed the rotors. They approached cautiously, as if closing on a surprised and potentially dangerous animal. When they had come close enough to talk over the whine of the idling engines, one of them greeted us. He wore a blaze-orange safety vest with reflectors, knee-high neoprene Chota boots, a sun hat with a flap that covered his neck, and sunglasses that covered deep wrinkles. A cigarette dangled impossibly from his lip. "You guys got a gun?" he asked, his lips parting on the word "gun," revealing tobacco-stained teeth.

"Nope," I said. "We've got bear mace."

His eyes narrowed behind the mirrored lenses. After a pause he went on. "We carry shotguns for grizzlies," he said, the cigarette dancing up and down. He kept one hand on his hip while the other gestured lazily as he talked. He looked around at us slowly, inventorying our boats and the six of us. "We thought you were a group of girls," he said, surprise and disappointment in his voice. His partner had been listening in silence, almost unnoticeable next to him. Now he laughed and shifted, still watching us.

"Well," I said, unsure what to say next, "we're not."

Cigarette chuckled, leaning forward onto his knee. "No, you're certainly not," he said.

We didn't mention our female compatriots paddling ninety miles to the east—certainly within range of the four-seater.

After a bit, the pilot got out of the helicopter, the rotors still whirring. He was dressed in blue coveralls, with a flight helmet and aviator sunglasses. In his hand, he held a huge Sherlock Holmes–style meerschaum pipe.

Cigarette told us about his work with the helicopter for the mines, and we told him about Hommes. The conversation naturally progressed to fishing. Dubawnt Lake, they told us, was nearly covered in ice, with only a couple of hundred meters of open water near shore. We could only hope that it would clear before we got there, or we would be in for a tough time. The enormous lake, nearly covered in ice, would be a huge obstacle.

The men told us they were conducting a land and geological survey for mining companies. They would fly about, all over the tundra, landing at refueling stockpiles placed in strategic remote locations.

Before leaving, Cigarette pulled a camera from his pocket and took a photo of our group as we sat in our three canoes. Then the three men returned to the helicopter. The rotors spooled their way back up to flying speed, and the craft lifted off the taiga, blasting us again with rotor wash and engine noise before arcing up and back over the lake. After a minute, silence returned. It was possible we might see them again in a month, when we paddled into Baker Lake. In the beginning of August, when we got there, they would be done with their survey work, and we would have put in several hundred more miles of paddling, including Dubawnt Lake and the Kunwak and Kazan Rivers.

The helicopter faded to a dot in the distance. We were alone again. The thump of the rotor, the whine of the engine, the uncomfortable conversation, the three odd men—all were gone. It was a brief encounter, but all six of us felt unsettled by it.

It wasn't until after one a.m. that we finally stopped and made camp in the twilight. We'd paddled twenty-four miles. We slept for nearly twelve hours and didn't get on the water until three thirty the next afternoon.

Our canoes slid silently through a mirrored sky, across miles of open water, toward a distant flickering mirage. We eventually reached it—an island piled high with sheets of spiky ice. Dan and Auggie landed there and pulled their boat far up on shore. Seconds later, we heard a crack as the canoe jolted loose of its beaching, slid down the ice, and splashed into the water like a newly christened ship. My heart leaped to my throat as we all watched one of our lifelines slip out of our control. Mike and I paddled frantically toward it. Every bit of training had me flinching at even the temporary loss of a boat. To see that cache of food and supplies slip into the seemingly endless expanse of water reminded me of the fragile line between being prepared and suddenly being in dire straits. I imagined a scenario of a swamped or lost canoe holding the main set of maps, a third of the food, and other critically essential gear, and shuddered.

As Mike and I raced to recover the boat, I recalled an incident from a year before, on an island a thousand miles away. We had made camp, our tent pitched and staked on the bald windward shore. A storm had built so slowly that we hardly realized it had whipped into a fury—until the staked-down tent began to slide. It took all six of us, lying like ballast against the fabric, gripping the seams and holding the swaying poles, to keep the tent from blowing away. An hour into the storm, the brass hooks holding the rainfly bent straight and snapped free, their cords twanging. Darin and I bolted outside, the rest of the guys still holding it down from inside. Sideways rain stung my face. We wrangled the fly and clamped the hooks shut with pliers. Trees groaned downwind. With each gust, their root beds rose and fell like the gills of a gasping fish.

Watching the trees, I ran through camp, shielding my face against the rain and howling wind to inventory our gear. The packs were secure. The tent was being dealt with. I checked our dry dock on the far shore. The first two canoes were there, unmoved between quivering juniper bushes. In place of the third boat was an empty patch of scraped moss. I ducked and panned, scanning the water and shore through the storm. The boat had blown away. Our canoe was gone.

I ran back to the tent and shared the bad news. There was nothing we could do until the weather passed. The tempest got louder. The tent poles bent into hoops around us while, downwind, trees pulled free of their moorings and toppled to the ground.

The storm finally moved on, and we crawled out of our collapsed tent to begin the search for the missing canoe. After half an hour, we found it a quarter mile from the island. It had submarined in a shallow bay, completely submerged and barely visible until the boats were right over it. It was intact and upright. The guys poured the water out and paddled back to camp. We'd been lucky.

I pushed the daydream aside as Mike and I reached our drifting boat. We tied a lead to it and towed it back to the island. It was a solemn reminder. *Keep your boats in check.*

When evening came, we slipped off the lake and back onto the river. We skipped dinner and pushed on through several sets as the river carved down into a canyon. It was getting darker. The dimness and the dark canyon walls melded shadow and light, muddying contrast and making the features harder to see. My energy waned.

The last set proved technical, and Mike and I rested in an eddy before tackling it. We were surrounded by bushes that rose from the river, their branches obscuring all but the heads and shoulders of the guys in the other two canoes. The river flowed under the stunted canopy as if through a flooded forest, the water barely visible through the foliage. The other two boats had already completed the upstream ferry through the shrubs, then made it across the main current. It wasn't complicated, but the river was strong, and the bushes and dim light made me nervous. We steeled ourselves.

Starting the forward ferry, we slid sideways past the undergrowth and approached the main sluice of the river. At the last moment, I turned the boat and we rushed to break out of the eddy. We popped into the current. But something was wrong. We didn't have enough exit velocity. The river could try to swing us. Worse, I'd made our angle too wide. It *would* swing us. In front of me, Mike pulled hard at his paddle, but the canyon spun around him. The next instant, we were facing the far shore, where the two other boats had already pulled out of the water. We were perpendicular to the river. The current swept the bottom of the boat downstream, twisting it with us inside. Our upstream gunwale dipped underwater, and the river poured in. *Shit.*

Acting on instinct, Mike pushed his paddle against the river while I pulled with mine. Our knees pressed against the boat and we twisted hard. The upstream gunwale rose out of the water, and we righted. It had happened in the blink of an eye. The boat sloshed clumsily with water weight. We worked to swing the bow back upstream. We were still in the thick of the current, still had to finish the crossing. Carefully now, I adjusted the angle, and the canoe edged toward the far shore. The water inside both delayed and magnified our momentum. We paddled hard and limped along, completing the ferry at last and landing on the rocks of the far shore.

Dan watched as we bailed our boat. He was surprised. We'd made a simple mistake that almost caused a terrible swamping.

Downstream, the river only grew in size and strength, carving a deep canyon as it wound several more miles before twisting out of sight. If we had gone over, we would have been swept into a deep, seemingly endless set of rapids. Mike and I were shaken.

After a few minutes, we reluctantly got back on the river, stopping at the next bend to scout. The sun was well below the horizon now, and everything played out in the slow blur of twilight. From the top of a rise, I looked down at rows of standing waves, cresting foam, and deep troughs. Beyond, long sections of flat water rushed downstream like a dark waterslide. Interspersed among these, huge boulders parted the water like the prows of sunken ships. The more we scouted, the more nervous I got. Dan wanted to move. He was getting us ready to paddle, to keep on adding miles and get out of this canyon.

"Dan," I said, "I don't feel comfortable running these." I think some of the other guys felt the same way and talked to him, too. Dan had wanted to run them and was disappointed that we didn't, but he listened and quietly acquiesced. We made camp, hauling our gear over an enormous lichen-covered boulder field to a plateau above.

Before climbing into the tent, Dan turned to me. "If you didn't feel up to it," he said, "it's better we waited."

In the tent, I laid out my sleeping pad and folded my jacket into my shirt for a pillow. I unstuffed my sleeping bag and wrote in my journal. I'd started to hear the slow, rhythmic breath of sleep from my tent mates when I heard a shuffling outside. It wasn't one of the guys—everyone was already in their sleeping bags. I held my breath, my ears buzzing to hear more, as I eased toward the mesh of no-see-um netting. Cautiously, I looked through the screen. There was an animal, not ten yards away. It stood on slender legs. A huge pair of curving antlers bobbed as it grazed and swung in a wide arc when it raised its head to look in my direction. A blond collar of fur wrapped like a shawl around its chest. It was a caribou, the first I had ever seen. It lowered its head and continued grazing. I watched until it had disappeared over the ridge.

Looking at the big rumbling set in the light of morning, I was glad we hadn't tried to run it the evening before. The Dubawnt boomed as it rushed past, flying down the wide canyon at twenty miles per hour in some places. I was happy to have Auggie as my partner. We paddled well together in big water. One by one, our boats bounced across the swift water, ferrying back and forth past the enormous boulders, from eddy to eddy, the river in a constant churn. Finally, we paddled out of the canyon's depths into a calm, predictable flow.

Downriver, we pulled over and climbed a hill covered in quartz boulders, for a lunch of ramen. The soup was warm and rich and reminded me of home. I relished every bite and felt content. I was happy to be past the deep canyon, past the rapids that had nearly swamped us the night before. The rest of the group must have felt the same way, most of them lying back on the moss, hats over their eyes. I sat quietly, breathing the clean subarctic air in deeply while I looked at the slope of quartz boulders and the endless tundra beyond.

Eventually, we gathered our things and returned to the waiting boats. The river carried us into another canyon, and we spotted the waves and steep drop of rapids. We pulled over to scout. The view from shore made us all nervous. The river was powerful, walled in by sheer cliffs, with obstacle after obstacle materializing in the water before us. First, we would run a slalom of huge recyclers. After the fourth, the river rolled in enormous standing waves six to eight feet from top to trough. Intimidating but navigable. We marched back to the boats.

Auggie and I got in and started down, weaving around the four recyclers before coming to the standing waves. They loomed like great, dark gunmetal-gray berms, the water roaring up and over them as the smooth crests held fast.

We approached cautiously, the tumult of flowing water roaring and splashing around us. If we came at them straight on, the length of the boat would bury either the bow or the stern, so we skewed our angle to almost forty-five degrees. The bow rose up over the first crest. As soon as the boat slipped over the wave, we lurched down the far side. We were canoeing steeply downhill. Auggie and I balanced our craft against the slope, digging hard at the water to keep the craft stable. The top of the wave rose to eye level, then over our heads. We flew down into the trough and were surrounded by water.

The force of the river, combined with our own momentum, pushed us right up the other side. For a brief moment at the top, it was bright again and the boat felt light. In that instant, I could see the river's meandering flow downstream and the tall, craggy shoreline. As soon as we reached the top, we slid down the other side of the wave and were again walled in by dark water. I imagined being in the hand of a giant and hoped it wouldn't close into a fist. Rising and falling, sliding and holding, we bounced up and over the rest of the waves. It felt like the final undulating slopes of a roller coaster before the track flattens out at the end. It was at once terrifying and exhilarating.

Pushing late into the evening, we finally made camp. By the time we were headed to bed, the dark clouds we saw earlier had turned into rain. It continued all night. In the morning, it was replaced by a driving wind.

On the river, we fought wind and current, pushed this way and that by the two competing forces. After only a few sets, tempers were flaring and guys were yelling at the wind and water. The sky was dark, turning the water steely gray and casting looming rock formations in dark shadow.

Dropping down a cold set, we scanned the horizon for an inukshuk. Soon the river would drop into a maelstrom that was beyond our abilities—a set of rapids that we would have to portage around. That inukshuk was the only heads-up for the coming portage. Dan knew it was coming—he had found this portage mentioned in several of the old trip journals he dug up in the dining-hall basement at Camp Menogyn.

I had been down in that dim, still space and seen them, piled on the worn wooden desk, stuffed in drawers and cubbies, stacked on shelves. Their spines faced out, names like *Hommes du Nord 1968*, and *Waputiks 2003* written in pen. There were hundreds—records of adventure, each a time capsule of wild North America. These were primary documents, written firsthand. Each held little gems of insight, scrawled by lamp- or firelight as guide or camper sat on the ground or lay in a tent at the end of a long, arduous day. The format of the journals was similar to the one we were writing on Hommes, drifting from the mundane to the barely believable, to the stoically underwritten. Dan had pored over these old journals. He had researched the route, then consulted these documents, found entries for the portages we would carry, the rapids we would run, the lakes we'd paddle, and the sites we'd see. Going from notebook to notebook, he had traced our route, jotting down information that might prove useful, hoping to learn from the mistakes and victories of those who came before. Those notes rode with us now, transcribed into our group journal and his own personal notebook. To me, they were like whispers from an old sage. *You'll come to a bend in the river twelve miles from the lake. Stay river right and look for the inukshuk ... It's easy to miss.*

The journals were right. We spied the small place marker from our canoes on the Dubawnt and pulled off the river, hoisted our gear, and hauled it over rock and scree to where the portage ended at a cliff. Rusty brown rocks like huge cubes jutted out of the water. Several giant steps down from where we stood with boats and packs was a landing, still several feet above the flowing water. This was our spot. We helped each other down, carefully lowering, loading, and launching one boat at a time and ferrying upstream across the swift current. Despite paddling as hard as I could, we slid slowly downstream, traversing the river sideways as we drifted past recyclers, pillows, and exposed rocks. As we pushed across the second half of the river, my arms began to tire. Upriver, the rapids roared. Below us, they foamed, blanching to a pale gray. Around us, the water rolled, bouncing and jostling us. We pushed on, finally reaching calm current near shore. A few more yards, and we were in the safety of an eddy with the rest of the group.

Downriver, the Dubawnt cut into a deep channel. Free of rocks and features, the river flew. We put miles behind us until we rounded a corner and were slammed by stiff winds that literally stopped us in midcurrent. It must have been blowing thirty miles per hour, with gusts far above that. We struggled against the sudden onslaught, the wind and current magnifying any movement of our boats. We had spilled out into a delta, the shore spreading far away to either side. In front of us was nothing but flat horizon and endless wind. My eyes watered.

Just at the edge of the windswept delta was the boundary between the Northwest Territories to the west, and Nunavut to the east. We had reached Dubawnt Lake, the biggest lake in Nunavut, and with it, our first view of the territory. We had known that Dubawnt Lake would be huge, and had given ourselves the better part of a week to traverse its great open expanse, but the view eclipsed my expectations. Wiping the tears and looking through the frigid wind, I noticed a white shimmer along the horizon. It stretched from shore to shore. A mirage? A reflection? After a moment, I realized what it was, and my breath caught. The man from the helicopter had been right. Dubawnt Lake was filled with ice.

On the 10th of August, 1893, the Tyrrell expedition found itself on Dubawnt Lake up against a great wall of ice and upon an apparently immovable field of the same terrible material. As far as they could judge, they had struck ramparts of the Arctic. There was none to tell them what lay beyond, and tho they supposed this frigid field was only a phenomenon, it was mighty discouraging to come across it and to have absolutely no knowledge of how far it extended.

—***Literary Digest*** (1913) on the 1893 J. B. Tyrrell expedition

CHAPTER FOUR

Dubawnt Lake: Days 14–16

Unable to push on in the wind, we made camp. By morning, it had calmed, and the sun shone brightly in blue skies amid welcoming clouds, but it was cold. The lake, with its expansive fields of ice, made the air crisp. We cleared our campsite, hiked down to the boats, and sped quickly down the delta. Soon, we had spilled out onto Dubawnt Lake. It was a slowly churning sluice of water and ice that stretched past the horizon. This would be a hard day of paddling.

For the first time on the trip, we saw a fish jump out of the water. Seeing it as a sign, I cast a heavy, shining spoon where it had leaped. Before long, I felt the rod pull back.

"Fish on!" I yelled. With the rod bent full over, I reeled in a long, stout lake trout, its spots shining in the sun.

At the same time, Darin, trolling from another boat, called out, "Fish on!"

We pulled over at a nearby island, and Auggie and I filleted the fish. Before long, we had four glistening fillets, skinless and boneless. They looked like salmon sashimi. We put them in our big pot filled with ice water to keep cold till dinner.

We hadn't been back on the water long before we started traversing ice. The bergs moved like tectonic plates, great sheets grinding against one another as if fighting for space on the surface. Occasionally, we had to get out of the canoes, ramming, pushing, and smashing our way through the ice. It wasn't good for the boats, but they would hold. They could take almost anything we could dish out—dragging across the ground, wrapping around a rock, dropping down a cliff—and keep on going. These boats were tough, built for capacity, maneuverability, and durability. Still, the rough treatment made me cringe.

All day, we toiled through the ice, weaving between the floes and through open spots of clear, frigid water where we could see the lake bottom and the fish as we glided past. In the evening, as we approached the island we were going to camp on, Auggie and I trolled again. Within minutes, I'd hooked a fish. I reeled in, but it kept pulling out the drag. I began to wonder. *Did I tie the line to the spool, or just wrap it around?* I couldn't remember. I tightened the drag as far as it went, and began making progress.

Fifteen minutes later, I saw the murky glow of my catch shimmer from the deep. It was a huge, steel-gray thing of muscle and teeth. When it saw the surface, the fish gained renewed vigor and twitched into motion, bursting downward once more. The drag whined as line stripped off the reel. I wrenched back against it. The moving shape grew again as it approached the surface; then a lake trout exploded into the air.

"Holy shit, it's a frickin' shark!" Jean yelled. It flapped violently, dousing us with ice water. Manhandling it out of the lake, we hefted it into the canoe. We pulled off to an island to process the fish, then pushed on to make camp.

By the time we reached camp with the fillets, all three tents were already up. The sun was low on the horizon and cast a deep golden hue over everything. Auggie spooned huge half moons of Crisco into the hot skillet, and we looked on hungrily as it sputtered and skated around in the heat. We pan-fried fillet after fillet in the hot oil and made french fries from scratch. The potatoes were hot and greasy, and the fish was succulent and rich. It was the best fish and fries I ever tasted. Thumbing the GPS, I made a new waypoint, as I did at each site. This one, I labeled "Fish and Chips."

The sun set with deep pinks and purples that lingered for hours as a sliver of moon rose over our tents. Auggie, Mike, and I stayed up to put things away and eat more fish while we watched the colors drift across the sky. Radiating out from our island, the lake was filled with thousands of shards of ice that sat like stars, motionless, glowing with the rich hues of sunset and sparkling in the moonlight. There was so much space, so much stillness, so much color.

The next morning was tough going, despite there being no wind. The ice was confining, and finding our way through the large floes and icebergs was exhausting. Up close, the bergs resembled the krill filter of a whale—parallel cylinders slowly rising and falling in our wake, hissing as the water poured in and strained back out. Scores of fish teased the surface with gentle ripples before breaking through with dorsal fin and tail like little sharks on the hunt. We weren't interested in any more of them, though, after stuffing ourselves the night before.

We quietly navigated a long, clear sliver of water between floes until, miles later, our channel finally closed in on itself, and we were stuck in the middle of an enormous ice floe. With no way to ram through or skirt it in the boats, we were forced to traverse the surface. But the floe was a patchwork of different-sized plates, varying from thick, solid white to blue, to muddy gray. The heavy, slushy gray gave way underfoot, opening up to dark, bone-chilling water. The blue ice was less predictable, like thin panes of glass on an old house. This ice could hold, but it might splinter under our weight and suddenly shatter. The white ice was best. Where it was solid enough to walk across, we donned the canoe harnesses we'd brought along for this exact purpose, towing the boats behind us like oxen skidding a log.

We kept close together, each of us holding our paddle parallel to the surface with both hands so that if we fell through, we would stop when the paddle spanned the hole, and we wouldn't slide under. The ice creaked and groaned underfoot. We grew increasingly worried about its integrity and eventually decided it was time to change tactics. To avoid falling into the water, we worked over the boats as much as possible. This involved both bowman and sternman bending over our respective ends, our center of gravity over the canoe, propelling the fully loaded boat across the ice, running awkwardly with bowed legs straddling the gunwales. It was hard work. We looked like some ungainly four-legged creature dragging its big red belly over the ice. It was loud and awkward, but it did the job. We called it "turtling."

Once we had gotten some momentum with the boat, we broke into a jog, then a sort of ungainly sprint. It was an awkward full-body workout. I was toying with the idea of running beside the boat for greater speed and ease, when the hard white ice cracked resolutely and my foot punched through to the dark waters below. Down I went, my body listing as my arms tried to react. Then suddenly, as icy water splashed and my knee plunged toward the lake, the hard gunwales of the boat slammed against my thighs and groin. I grunted, nearly kissing the forward deck plate, when my arms finally arrested my fall. Before I could even try to pull my leg out, the hole I'd just punched swept by, and my foot popped back out on its own as the boat's momentum dragged me forward. With the canoe still skidding as the sternman kept galloping, I pushed back up to my feet and turtled on. It happened more throughout the day. It was precarious, but it wasn't unsafe. It did, however, involve thin ice, and water just above the freezing point. As with everything on this trip, we were managing our risks. It was brutally tiring and lots of fun.

After ten hard-earned miles of turtling, we stopped for the night. The air off the lake was bitterly cold, but it made for fewer bugs. The ice moved on the water in silence, and in the open pools between the bergs, the raised fins of trout moved noiselessly, rising and falling like submarine periscopes.

The next day, we managed to paddle several miles before stopping at an island to swim off the ice. The sun warmed the air, but the water was as cold as ever. Standing in the water, preparing for the shock of the swim, I admired the bold blue sky, the bright blue-white ice, and the dark shadows of deep water. I felt present and alive. There were no distractions of home, no thoughts of how we came to be here or what might happen farther down Dubawnt Lake. My mind was as clear and pure as the crisp air around me. I let my body crumple into the ice water. Cold enveloped me. My skin stung. My face cramped, and I immediately had a brain freeze. In an instant, I lurched back up, gasping as my muscles seized. Then, my body relaxed, and I suddenly felt warmer than before.

As we paddled into camp after a calm seven-mile crossing, the air chilled. Water froze around us as we paddled, the boats cracking through the tiny film of ice as it formed. By the time we made camp, my motionless feet were going numb. We cooked a steaming stew to counter the cold. By nightfall, it was even colder. I was happy to climb into my cozy sleeping bag and cinch the hood down tight until only my nose was left exposed to the frosty air.

> Difficulties are just things to overcome, after all.
>
> —**Ernest Shackleton,** *The Heart of the Antarctic*

CHAPTER FIVE

The Final Push on Dubawnt Lake: Days 17–18

"Time to get up," Dan said the next morning. "We're iced in."

We poked our heads out the tent doors. Completely surrounding our canoes and the island where we had slept were dense rafts of ice and slush—frozen muck, spread out for miles in a still, ragged sheet.

We climbed to the top of the island to see how thoroughly we were surrounded, and from the pinnacle we could see that the island was entirely encased in ice but for one side. That was the only way back out onto the lake, via the north shore. We hiked back to camp. Not wanting to portage the shore of an island, we decided to scoot the loaded boats along the edge rather than carry them. Mike leaned into a canoe harness, and I ran with the boat behind him, guiding it around boulders and heaps of wind-stacked ice. After much effort, we finally reached the ice-free north side and launched the boats onto the open water we had seen from above. We whooped and hollered, having conquered the obstacle and escaped. We paddled freely, chatting happily as we went.

We paddled for miles before encountering more ice, and as we approached this next floe, we were flanked by enormous sheets that stretched out to either side. Pushing deeper into this channel, I noticed something out of place on the ice—something small, gray, and round. As we drew closer, I made out the gentle curve of a neck. It was a goose. Its head rested on the ice, the neck curving around its body. Despite the slump, it was standing, feet planted. It was motionless. Was it asleep? As we silently drifted closer, I realized that it was dead. I was suddenly filled with a sense of foreboding. The goose's feet had frozen to the ice, trapping it until it succumbed, killed by time and cold.

In the darkening gray of the day, we went deeper and deeper into the passage between the floes. Almost imperceptibly, it closed around us until we were surrounded by a mess of candlestick ice, bergs, slush, and frigid water. We made slow, grueling progress through the mass, paddling at first, then scooting, pushing, turtling. Bound by the ice, we pulled with our paddles and kicked at the bobbing mass with our feet. Eventually, our boats wouldn't move. We were in a vast white quagmire.

The ice would not support our weight, and it was too thick to push through with our paddles. This raft of candlestick ice was like a bundle of twigs, miles wide, all moving together with the water, falling apart and opening to the inky depths below when we applied any weight. We were stuck. Had to get unstuck. Lying across the boat, legs hanging over the side, I pushed out and back against the ice with my boot. When I pressed sideways at the candlesticks, they held. Groaning and pushing, we inched the boat forward. With Mike and me working together, kicking out at the ice with both legs, so that the canoe looked like some bizarre crocodile waddling across the ice, we made some progress. It was strenuous and slow going. Soon, I was exhausted, aching, and out of breath, splayed out across the gunwales, with little to show from our efforts. This wasn't going to work, either.

As I lay there panting, I tried to think around our predicament. We were locked in ice too thick to push or paddle our way through and too loose to bear our sustained weight. *What if it didn't have to sustain our weight?* I pushed down with my boot, and it stood there for a moment before plunging down into the dark water. The longer I held my boot there, the faster it sank. I tried slapping my foot against a fresh patch of ice. The ice barely moved. I tried it again, this time holding my boot there. For a moment, it held. Then the crystals sank, coming apart in the water like the splinters of a colorless firework. Maybe, it would hold us if we moved really fast. Not far from our boat, spots of whiter, more solid ice dotted the surface. We were close enough to one that I could just reach my foot out to it. Mike and I got into turtle position and began playing a widely spaced game of timed hopscotch, trying to land on and push off the white spots before they started to give. It was more efficient than our crocodile crawl, but the stakes were high. When we missed the spots or slowed down, the shards of slush gave way and we plunged into the water, only to stop as we slammed into the gunwales of the boat. We leaped and kicked our way across the floe, leaving a trail of sinking steps in our wake as we made our way to the mist-shrouded island emerging in the distance.

Exhausted, we hopped a small crevasse and finally pulled the boat onto a sheet of solid ice. At the far edge of the sheet, blue mountains of ice towered where the floe crushed up against shore. We scouted a path and muscled the canoes over. Feeling clumsy and drained, we slid ourselves and the boats and stumbled onto the damp but solid island, relieved to be finally on land and out of the frozen muck.

We had to scout our next move, so we walked up the island, fortifying ourselves with trail mix as we reached the high point. Surrounding all the island but the far side were more floes like the one we had just forged through. But off to that far shore, almost a mile away over ruddy, tussocked tundra, there was a way out. The only way off the island and out of the grip of the ice was the far shore. We would have to portage across.

As we stood in the cold, cloudy gloom of the windswept barrens, we looked into the distance, at the sunlit far shore and the open water behind it. Dan studied the ice on either side of that wide-open channel to freedom. He watched, tracking the slow movement of the most distant ice. While we were standing and reconnoitering, the winds shifted, picking up. He squinted, shielding his eyes from the reflected sun. I imagined him tracking enemy troop formations miles away, scheming, plotting attacks and counterattacks.

Peering into the shimmering white, Dan could see it now. The open lane we'd been watching was essentially a mirage. He pieced together the slow motion he'd been watching, and suddenly his eyes went wide.

"Guys, we need to move!" Dan said, startling us out of our malaise. He was serious. "Hurry it up," he said, already several long strides down the hill. "We need to finish this portage as fast as we can!" He had seen what we missed, though we'd been looking right at it. The two distant edges of ice were slowly converging, closing off our only escape route from the island.

We ran down behind him and loaded up for the first of two trips across the three-quarter-mile portage. The island was nothing but swampy tundra, with uneven terrain and maddening tussocks throughout. These bundles of grass rose straight up from the ground, blossoming outward at the top—spongy pylons whose sole purpose in life was to trip and foul us as we went. Portaging with the canoe, I accidentally stepped on one. It tipped over, nearly toppling me in the process. The next one succeeded when I stepped on it and fell to my knee. Somehow, I managed to keep the canoe on my shoulders, hauling myself back to my feet with many grunts and curses. As we marched, the wind picked up, whipping into a gale. The canoes were like sails. Turning the boat to shed the wind, I walked sideways, sidestepping over and around the tussocks. All the while, my knuckles were white as I held the boat tight to keep it from ripping off my shoulders and flinging across the tundra. Dan pushed us all the way, yelling over the wind. We could barely hear him. "Keep it up, guys!" We'd lean into our straps, into the wind, gripping the cold gunwales and pushing on harder than before. It was a race against the wind and the ice, which was closing faster now. I finally set the boat down at the end of the portage and wet my throat with water from my bottle. In midchug, Dan sent me back for the next load. I ran back.

By the time we were finishing the portage, the channel had nearly closed. Before I had even set my pack down, Dan and Jean were already several turns into the maze of the floe, trying to find the last open leads before the ice closed up completely.

With our boats loaded, we pushed off, giving chase behind Dan, with the third boat just behind us. Dark clouds swirled and the wind howled. We zigged and zagged, the ice shifting around us. At every turn, I looked to Dan's boat and traced the open leads back to us to find the route. We paddled the open leads and crossed over the solid ice between them, turtling at a jog, jumping from one solid white patch to another amid a sea of black candlestick ice. It felt like walking on water.

Halfway through the floe, Dan's boat had managed to slide through a narrow lead, but by the time we got there, it had slammed shut. We were too slow. The ice had cut us off. Dan's canoe still zigged and zagged through open channels as if navigating a tight stream, but we were stuck. Looking back down at our obstacle, at the solid seam, we tried to imagine another way around, but in front of our boat were two enormous sheets of ice. One piece was the area of a football field. The other was a quarter-mile square of solid ice. There was no going around it. In the distance was the shimmering image of Dan's boat as he paddled on through. Beyond him was more ice, speckled white plates strewn across the water. Beyond that was the mirage of open water, whipped by the wind into a deep blue. I looked back down at the vast white sheet in front of us. *I'm sick of this,* I thought. I was done turtling. My legs were tired, my thighs sore from falling repeatedly onto the gunwales.

I turned the options over in my mind before climbing out onto the ice. It felt as solid as the crust of a midwinter lake. Still, I knew that the plates weren't fused. When we had stood on small floes, the ice was quite buoyant, tilting and bobbing with our weight but rarely sinking down. On bigger sheets and bergs, even though the ice wouldn't give, it could slide along the surface like a canoe skimming across a lake. The two pieces in front of us were just huge versions of the plates we'd played with before, and with enough force exerted between them, they would part. I just had to figure out how to apply that force.

I sat down at the seam of the two floes. Beneath the crack, through the tiny fissure, I could see the dark shadow of the lake. I dug my heels into the raised lip at one edge and sank my hips down to catch the other edge, gripping the ice with my hands. Taking a deep breath, I started the leg press. The ice didn't move. Groaning, I kept working. The hair on my neck rose. I could sense momentum building. I kept pushing. Slowly, the seam started to give. My legs were shaking now. Gradually, like the doors on an enormous elevator, the crack below me began to open. As the momentum grew, the ice parted, until suddenly they were sliding apart without my help, and with increasing speed. Dark waters opened up beneath me, and the canoe bobbed free.

"Holy shit!" I crowed. "It worked!"

I leaped up from the growing chasm, scrambling to my feet as the ice slipped out from under me and hopping back into the boat. The ice crunched as the entire floe shifted. Knowing that it wouldn't stay open for long, Mike and I paddled hard, laughing as we went. The canoe accelerated down the still-widening channel, and Darin and Auggie were hard on our heels.

Halfway through, the crunching quieted, and the ice suddenly went silent. The sheets on either side lingered, motionless. Then they began to converge. "Hurry up!" we yelled back. "The lead is closing!" In a few hard strokes, our boat was out of danger, in the open pool beyond the channel. Darin and Auggie, paddling furiously, were halfway through the lead. It narrowed. The ice rang like wind chimes as its momentum grew and hundreds of feet of solid and candlestick ice settled toward them. It was just feet from their boat now. Darin and Auggie threw in one last hard stroke. Their canoe launched forward, and they lifted their paddles out of the water, the ice so close they couldn't paddle any more if they tried. I held my breath, waiting for them to pop up like a cork. Everything was coming together. Then, as if by magic, their canoe slipped past the last lip of ice, and the lead slammed shut behind them with a slosh of water and a heavy, stony thud.

They'd made it through, free and clear.

"Holy shit!" Darin said, still sliding, an uncontainable grin on his face. "That was awesome!"

Beyond, the floe opened up, and we finally caught up with Dan and Jean. Before reaching open water, we came upon strange, otherworldly ice formations. It formed a shelf as clear as glass, two feet under the water's surface. It must have been only inches thick. The ice felt solid yet brittle, like walking on windows, and had a tinny, flat resonance like a glass baking dish. It didn't inspire much confidence. We walked carefully, sloshing through calf-deep water. Farther on, it was punctuated with clean holes, like pockmarked limestone, that opened straight to the belly of the lake. Through them, I could see the boulder-strewn lake bottom some forty feet down. The deep water glowed a rich blue. But for the penetrating cold and pervasive ice, it could have been the Mediterranean. I cautiously appreciated the haunting beauty and, feeling vulnerable, stepped as lightly as I could. It felt as if we were suspended between worlds.

From there, the ice pack gradually dispersed until we finally reached open water. With the air just below freezing, we bundled up against a bitter wind for the final paddle of the day.

In the morning, still weary from the long ordeal of the day before, we slowly filtered out of our tents before continuing north across Dubawnt Lake. The sky wept all day while the temperature dropped and the tempestuous north wind whipped the gray waters into enormous frothy waves. The lake seemed to boil as we made one large crossing after another over the rolling water. We rose up over crest after crest, the boat slamming down and spraying Auggie, in the bow. It was slow going, but we moved steadily forward.

On our last and longest crossing, the waves were so big that when another boat entered the trough, it almost completely disappeared from view. Timing my paddle strokes to some fixed cadence didn't work. When I began reacting to the natural movement of the water, it became a dance.

We paddled like this for hours. It felt like rolling on the backs of great humpback whales.

After that crossing, we stopped at the first spit of land we found, a large peninsula. The ground seemed to glow green with a carpet of moss that rose sharply not far from shore into abrupt rock-strewn slopes. We walked to warm up and get blood flow back into our leaden legs and refueled from an enormous bag of hard candy.

The lake calmed while we rested, but when we returned to the boats it whipped back into the tempest we'd been fighting all day. We paddled hard, finally making camp at midnight on the only livable island for miles. The island didn't rise more than a few feet above water and looked more like a debris field from a volcanic explosion than a place we could sleep for a night. The ground, where it was even visible, was covered with tufts of dry grass, like a bird's nest under huge granite eggs. We pitched our tents on two six-foot squares that were relatively rock free. One tent pad sloped gently. The other dropped more than a foot from one side to the other, sagging in the middle and rolling even deeper toward one end.

In twilight, we set up camp and started unpacking our food among the boulders. Parting out and inventorying ingredients from the huge blue barrel, we discovered that our food situation was even worse than we thought. We were drastically low on pasta—not just a portion here or there as we'd thought, but multiple meals short. The six of us stood quietly, looking down at the meager pile of spaghetti, trying to imagine how it had come to be so small while our hunger was so prodigious. But the answer was easy to see. We had made mistakes in our measuring. A short scoop here and a missed scoop there had added up to a sizable deficit. We made our modest meal in the darkening gray haze of night, the food shortage looming in everyone's mind. Fishing had been reliable, but who knew whether it would continue to be. We would have to start rationing.

We drew straws to see who would sleep in which tent. One by one, we pulled blades of grass from a bundle in Dan's fist and learned our fate for the night. I compared my grass with that of the other guys. I was holding one of the three short ones, along with Jean and Darin. It would be an uncomfortable night.

Neither tent pad was good, but as we laid out our sleeping pads, we found that ours dropped even farther than I'd thought—several feet from end to end, with buried boulders sprinkled throughout. In the middle of the tent was a low spot that ran down the slope lengthwise, like a gutter. That was where I slept. As terrible as the spot was, we were exhausted, and as soon as we were done complaining about it, we fell asleep.

> I want to know it all, possess it all, embrace the entire scene intimately, deeply, totally.
>
> —**Edward Abbey,** *Desert Solitaire*

CHAPTER SIX

Dubawnt Canyon: Days 19–20

Through the night, Jean and Darin slid toward me while all three of us slid down, our feet scrunching together at the bottom. The small island was even more desolate in the wan light of morning, and despite a strong wind that blew mist and rain, we were only too happy to stow our tents and leave. We could handle this weather.

We paddled north, toward the end of Dubawnt Lake. Hours later, squinting through the mist, we sighted the far shore, and the lake's northern terminus. We had crossed the largest lake in Nunavut. I heaved a deep sigh of relief. After spending a week crossing this single lake—all the paddling, ice, and effort—we finally felt the old pull of the current.

Before we dropped down into the river, the calm waters of the lake began to shift in an almost unnatural way. I felt as if we were sliding across oil. The water seemed to stretch then tear apart as up-currents broke the surface in flat, swirling geysers. We could feel it. We were approaching Dubawnt Canyon and its mile-long stretch of class III and IV rapids.

Slipping into the flow, Dan and I slid past rocks and pillows in narrow channels, still fighting a stiff crosswind. We hugged the shore. According to the journals, we were on the lookout for a "big cairn, you can't miss it," which signaled the start of our portage. We scanned the barren, rocky shore and the steep bank above it as we avoided more obstacles downriver.

Farther down, we spotted a three-foot pile of stones poking up from the rocky hilltop. It would have been easy to miss, although it also would have been foolish not to get off the river to scout from there. Downstream, the river whipped into white and blue froth in a rumbling frenzy of rock and waves before dropping after a distant bend.

We hauled our gear and two fresh trout up the steep scree of the riverbank to the main plateau. I filleted the fish while the others prepared to haul the boats. Up on the exposed ground, the wind grew to a gale. It was difficult even standing, let alone carrying a seventy-five-pound canoe that wanted to whip like an untied sail. Not wanting to risk injury from the boat flying off our shoulders, we opted to tow the boats instead of carry them. We ate the fish, then attached the harnesses to the canoes and began slogging. I hefted one of the packs and leaned forward against the weight. I caught up with Darin, leaning steeply into his canoe harness, and fell in behind him so we could switch off when he needed the break. It was a relief not to worry about the boat threatening to blow off your shoulders, but the pain of dragging them was almost worse. Almost.

Soon, we lost ourselves in the labor and exertion. We plodded on. With Darin finding our way across the emptiness, I was left to my wandering thoughts. I took in the surroundings: the muted colors, the many textures, the smooth waves of clouds as they drifted past, the distant snaking forms of mighty eskers, the dark scar of Dubawnt Canyon where it cut through the smooth roll of tundra.

Halfway across the portage, we climbed a gradual slope where the ground shifted from moss, flowers, and bushes into a uniform gravel bed covered with dark speckles of lichen. The smooth rock field was a quarter-mile long. Just north of the field, the steep canyon dropped away. It was picturesque, with dark-red cliffs descending to fast water with huge waves. We set down our gear at a pile the others had started. This rock field would be our home for two nights. Tomorrow was our first layover day.

Returning with our final load, we pitched our sleeping tents near the canyon's edge, with a view down the nearby cliff. From there, we could clearly see the direction we had come from, the jagged scar of the canyon, and the endless tundra beyond—a breathtaking spot. The plateau was exposed, though, and Dan and I set up a windbreak around the kitchen area, weighing the canoes down with food barrels and equipment.

Nestled into my sleeping bag that night, I thought about the next day. It would be our halfway point. We'd been paddling and portaging for nineteen days straight.

I awoke filled with anticipation and was shocked to find that it was one p.m. Mike was flipping pancakes, and near the stove was a steaming pot of syrup made from butter, brown sugar, and a hint of spices. Our well-deserved layover day was off to a great start.

After breakfast, I napped, but when Jean and Darin returned with broken rods and stories of monster lake trout, I climbed out of the tent, eager to try my luck at this magical fishing hole. Casting from a rocky beach at the bottom of a cliff beside the raging river, it wasn't long before I'd hooked a few fish of my own. The fish weren't biting hard, though, and most spat the hook when they were almost within net's reach, as if taunting me. With the third fish, the knot I had tied slipped and the fish swam away, one of my best lures still shimmering in its mouth.

Defeated, I climbed back up the cliff to camp. Taking extra care, I tied a new leader to the heavy line and clipped a big copper spoon to the end. Within a few casts, I had hooked a trout at least three feet long. After a prolonged fight bringing it to shore, we landed it in a net. It was the perfect specimen of a lake trout, with sleek lines and beautiful patterning. I hauled it carefully up the cliff.

By the time I reached the plateau, my back and arms were cramped and burning. I hiked across the flat to the fish-processing spot and was soon ready to skin the two fillets. Pinching a patch of skin at the narrow end, I slid the blade, bending the knife as close to the skin as possible without cutting it. On the second fillet, I began the same way, bending the flexible blade for another efficient slice. Halfway through, my hand jerked. There was a pop. I stabbed the ground. Startled, I froze, thought through where the blade was. Had it cut me? I couldn't feel a cut, but that didn't mean it hadn't. I turned the hilt, looking at my hand, at the contoured plastic, the knife length adjustment lever, the sharp blade where it protruded from the handle, and an abrupt break an inch away. The other end of the blade sat motionless in the meat. The knife had snapped. Its core, the hidden steel we should never see, glinted in the light.

I finished my work with the now awkward blade and retreated to the tent. I felt defeated by the loss. It had been a good knife. Lying on my back, I slipped into a daze, the sounds of the river and of the rest of the guys' activity wafting over me. I listened to Jean cooking his trout in the kitchen, a far-off conversation between Mike and Darin, the steady wind on the tent, and the distant roar of the river, all the sounds mingling in the air.

I was pulled from my reverie by a call to the kitchen for a snack of trout. I took a piece, was underwhelmed by its underdoneness, and left to walk by myself up the river.

Now was as good a time as any to take stock of myself, to be alone with my thoughts in this remotest of places. If everything went right, it was twenty days before we would paddle into Baker Lake, that tiny outpost at the end of our route. It was still a blank picture in my mind, but a concrete goal at the end of our six-hundred-mile journey. Much had happened up to this point, and I could sense a growing momentum, as if we were building up to some unknown culmination.

Reaching the end of the campsite plateau, I turned and took in the scene. All around me were endless fields of sage, yellow, and brown undergrowth, punctuated by scrubby bright-green bushes. At the scar of Dubawnt Canyon, the earth puckered down along the river. In some spots I could see the rumbling waters, but much of the canyon was visible only as scoured stone in a sea of rolling tundra. Beyond the unseen mouth of the canyon, the glacier-blue of Grant Lake spread for miles. From where I stood, our camp looked miniature, again like a diorama, the bright-yellow jackets and red canoes like shiny bits of carefully carved and painted wood. I could see the guys talking and laughing, but I could no longer hear them.

Turning back upstream, I made my way along the canyon's edge until I finally reached a jog in the river I'd seen from below. Dark, jagged granite cliffs dropped down to the tumult of water. Punctuating the deep hues of the granite were flat patches of pale-green lichen, while farther down the cliff were rich blooms like rusty splotches of cinnabar. At the bottom of the cliff was the frothing flow of the Dubawnt, in places pale blue, like long-weathered copper. Camp was out of sight. I heard only the rush of the water and the rustle of wind as it dwindled and then gusted back to life.

I sat for several minutes, taking photographs and settling into my private space. My mind wandered to daydreams about the future. I imagined our homecoming and warm greetings by friendly faces. I pictured explaining all this to my classmates in my senior year of high school. My thoughts soon shifted to more tangible worries. *What happens if we swamp on a set? What if a bear takes our food? What if a bear attacks one of* us*?*

I had had dreams about that while sleeping, and fears of it while awake. I always had my knives with me in those dreams and easily was faster than the animal. I would extend the fillet knife to its full length and trick the bear into biting down on my hand, then turn the fillet knife straight up so the force of the bite pushed the blade up through the animal's palate and brain. At that point, like with most dreams involving life and death, at the moment of greatest potential harm, the image would disappear, bursting into a cloud of mist. With my fillet knife broken now, that aspect of the dream was gone; the bear would not die. I'd had dreams before, though, that had not stopped. Dreams where I died, laid open by claws and crushed by teeth.

I quelled the thought and looked again at the rugged beauty of stone and river and tiny tundra blossoms before me, photographing them before packing up my camera and making my way back to the campsite.

I climbed up to the flats and stowed some of my gear near the tents, meeting Mike nearby. As he and I talked and the seagulls far behind us picked at our dead fish, we looked across the raging Dubawnt to where the tundra stretched out into the distance. On the horizon was the pastel orange glow of sunlight breaking through clouds along the edge of the earth. The view was of incomprehensible scale. The tundra is far from empty, and being in it, I appreciated its immensity, its unique beauty. This was an endless open wilderness.

I thought back to when I first imagined this place, when I met Dan, Mike, and Jean in the musty basement of the University of Minnesota's map library earlier in the year. I had just received a scholarship from Menogyn to help pay for the trip. Going on Hommes was concrete, but like a distant shoreline shrouded in fog, I couldn't picture it. At that point, I still had to finish my junior year of high school, go to friends' graduation parties, fly to the southern hemisphere to scout Peru with my family for one of the upcoming study-abroad trips that my dad would be teaching. Hommes would start after all that, not two full days after my return from Peru. *One thing at a time*, I had told myself. If I tried to imagine it too far out, my head would spin.

We descended into the basement of the library. The low ceilings, the smell of old papers and books, the whole purpose of our meeting, put me in a mood of exploration. We entered a room bordered with enormous flat file cabinets holding countless maps in great thin drawers. Dan had already pulled great sheets from some of the files and had some laid out on a table. We were discussing our route options, but he had a favorite: the Dubawnt, Kunwak, and Kazan Rivers. The route stretched all the way from the edge of the Northwest Territories up to Baker Lake. A bush flight would drop us off on the Dubawnt River in the great field of green on the map that signified the taiga forest. As the river meandered north and east, it passed countless bodies of water until it entered Nunavut at Dubawnt Lake. Halfway across the lake, dividing it north and south, the green field of trees stopped.

"It's the tree line," Dan said. "From there, it's nothing but tundra." I had looked at the map and imagined myself months into the future, in a canoe. I pictured launching from a stand of swampy, lichen-covered black spruces into a Dubawnt Lake wrapped in fog. We would paddle through the void of thick, moist air until, partway through, the fog would clear, revealing a far shore barren of trees. It mystified me.

"What do you guys think?" Dan had asked the three of us as we pored over the map.

The memory faded as I gazed out at the steady slide of clouds and the surging river in Dubawnt Canyon. There were no trees. We had crossed that line on the map, though it wasn't what I'd imagined. Not in a bad way, but it was just different. I'd been naive to think that the trees would be there one mile and gone the next. They had gradually dwindled, grown smaller and fewer until they vanished entirely. Now that I thought about it, I hadn't even noticed when they disappeared. It was one of the first daydreams I'd had about this place, something that had replayed in my mind before I came here, and I hadn't even noticed when it happened. Maybe I'd been waiting for the fog. Maybe the trees had just disappeared so slowly that we didn't even realize they were gone. I told myself to pay more attention, not to forget to see and listen and feel.

I took in the tundra in front of me. It was like a painting, a grand landscape. It looked planned, dutifully brushed, and carefully placed. As I looked on, though, I realized that it was no longer still. There was motion on the tundra. Something was moving. It was light tan, big, and surprisingly fast. It rolled fluidly over obstacles, as if it had been doing it for a lifetime. It was headed in the direction of the canyon. It was headed straight for our camp. As I stared, I realized what we were watching. My pulse quickened. My stomach tightened. Without turning, I pointed.

Mike yelled, "Bear!"

Dan, Darin, Auggie, and Jean all turned. Our camp was suddenly on high alert. Wide eyes scanned the horizon. They found it, slowly registering that the threat was on the other side of the canyon, and we gradually eased into cautious, captivated observation.

When Mike yelled, he alerted the bear, too. It stopped, startled from some inner reverie with the sudden announcement that it was not alone, then turned right around to head in the other direction. Its retreat was a methodical gallop, much faster than its original pace. I imagined the sounds the bear made, the rhythmic thumps as its wide paws hit the tundra, and the steady breaths as it labored. The bear seemed calculated in its speed, knowing that it wanted to get away but that it had a long way to go. It was pacing itself. Even from this distance, it looked as if the creature had a huge untapped reserve of energy and speed behind its every move. Looking over the canyon and far across the tundra, I sensed its intelligence, recognized its intention.

By this time, Dan, Darin, Auggie, and Jean were standing near us, and we all had our cameras and binoculars to get a closer look at our first grizzly. The huge set of white water roared from the chasm in front of us, and we were now thankful for the lack of any good crossings for a mile in either direction. We stared across the great canyon at the shrinking sandy shape. Grasses shook with the wind, small bushes swayed, and the river still raged, but to the six of us, there was only the bear, moving fast across the still tundra until it finally disappeared in the distance.

Slowly, we each turned away, looking back every so often.

My heart was still thumping in my ears. I turned to Dan. "That's the only way I want to see a grizzly bear in the wild," I said, pointing, my voice a little shaky. "Far away, on the other side of a canyon like that."

According to Canadian canoeing legend Bob O'Hara, whom Dan had consulted before the trip, we were lucky even to have *seen* a grizzly. I know I felt lucky. We'd gotten to see one, and safely. But a tinge of worry lingered in my mind. *Luck.* I reminded myself. *We were lucky to have seen it. This is a good thing.*

It wasn't likely that the bear could cross the canyon; the water was just too intense. It could meander up- or downstream and cross the placid waters of the lakes at either end, but that didn't seem likely, either. In either case, we looked at our site. As always, the cooking area was a good distance from our tents, as was our food storage. Our filleting area would arguably be the most attractive to a bear, but it was far away from everything, as it should be. We weren't miles away, as we had been most nights, when we stopped to fillet long ahead of making camp, but we were well set up, our camp laid out as it was supposed to be. We reviewed our inventory of equipment, talked through the protocol for bear mace and the bear poppers—deafeningly loud noise deterrents—and the unlikely event of a bear coming into camp. We'd been trained at Menogyn, but we suddenly all wanted the refresher course. Dan turned the can of mace in his hand, pointing out the glow-in-the-dark safety wedge and the flap of a trigger underneath and reminding us that it sprayed thirty feet for about seven seconds before it ran out.

"If you run into one," Dan reminded us, "talk to it in a calm voice. 'Hey, bear, it's okay, bear.'"

It made my hair stand on end.

That night, Dan put the bear spray in the vestibule of each tent and placed the bear bangers under his pillow. We had always been mindful, but now we were on alert.

As I lay in my sleeping bag, I thought about the bear. I pictured it rolling across the tundra. *Where was it headed?* Before we'd startled it and turned it around, it had been on a straight course from the distant horizon to our camp. My skin tingled as I thought about it. Was it coincidence? No, I decided, it was not. This was more than chance. How else would it have been walking directly toward us? I rolled over and focused on the class IV moat between us and the bear. *We are prepared as best we can be.* Besides, bears were rare here, we'd been told, and we were lucky to have seen one.

> When you put your hand in a flowing stream, you touch the last that has gone before and the first of what is still to come.
>
> **—Leonardo da Vinci**

CHAPTER SEVEN

Ebb and Flow: Days 21–23

When I awoke the next morning, I could sense a change. We were past the halfway point and were now counting down the days. None of us was eager for the trip to end, but we all missed aspects of our lives back home. I had always felt that a journey was incomplete without a homecoming, the anticipation of coming home being an integral part. We would be coming back to Menogyn having completed Hommes, the longest, remotest men's canoeing trip the camp offered. We would be welcomed with a feast and a hot cleansing sauna, and we'd tell stories to friends and family over a campfire before putting our symbolic red paddle on the wall, to join the history of the long trips.

I imagined it, then shook the image from my mind. *Don't think about going home. It'll be over before you know it; don't wish for it.* The purpose of the journey wasn't to think about home or how far I was from it, or what it would be like when I got there. The purpose of a journey is to experience those things that can't be explained and to forge the memories that will never be forgotten, the ones that change you forever. Starting now, we'd been on trail for more time than we had left. I was suddenly worried that it all would end too soon. I quelled the thought and stuffed my sleeping bag.

Once we'd packed our gear and struck camp, we waited for Dan's cinnamon rolls to finish baking. Finally, he picked the silver cover off the steaming oven to reveal a glistening pinwheel of deliciously sugary rolls. The sticky sweetness of the filling was heavenly, no doubt enhanced by our fat- and sugar-deprived diet the past three weeks on trail. The rolls didn't last long.

For this leg of the portage, Mike, Darin, and I took the canoes, dragging them again down the winding canyon and over the tundra to Grant Lake. The portage ended at a short cliff at the end of Dubawnt Canyon, where we carefully lowered the gear and canoes, loading them one at a time. An unimaginable amount of water roared past. Standing at the top and looking down at the relentless power of the river, I suddenly felt overcome, my legs shaky. I took a long steadying breath. It had happened several times on the trip, this feeling of being overwhelmed and overawed by the river—once when Mike and I nearly swamped; once gazing from the top of a ridge, where the river seemed to go on to infinity; and now, from the end of this portage. I let the feeling pass and climbed down to the last waiting boat.

We fought the chop from the bottom of the canyon and finally reached the calm waters of Grant Lake, already exhausted. Switching back to the rhythms of lake travel, we headed northeast toward the far bay, where we would again pick up the river. Paddling along, we noticed something moving on the left shore. It was light-colored, moving steadily, easily. It was somehow familiar. As we got closer, my adrenaline spiked. We were looking at another grizzly bear. Our second in as many days, or was it the same one? Its light tan, almost gray coloration went against my perception of what they looked like. It didn't pay much attention to us and went about its business. We paddled on, cautiously looking over our shoulders at the wandering creature.

We paddled downriver through swift and calm water to the next lake and began fishing. Before long, Darin had hooked a big one. These lakes were clear, but they were deep, dark, and mysterious, and seeing the fish rise was like watching something from another world. It grew and grew. Then, seconds before Darin could get a net under it, the trout threw the hook. The lure shot out of the water. Suddenly free, the fish turned and charged the boat, slamming into our canoe with an audible thud before disappearing below us. We sat stunned. Rammed by our own white whale.

Indignant, Darin sent the spoon out again. He spun the reel, and the lure came back to the boat too easily, empty. The last few feet, it wriggled within plain sight, as if to taunt us. Then the huge trout loomed out of the shadows and struck the lure. The fight was on once more.

The drag on the reel whined as line played out, and the bow of our boat turned toward the diving fish. Darin tightened the drag and slowly reeled it in until it was finally back to where it had started, just feet from the boat. Darin awkwardly pulled on the rod with one hand and reached down with the other, getting a grip around the slimy tail, his arms spread-eagled. He had to stand up to get it in the boat, then held the enormous trout triumphantly in both hands, wearing a smirk of surprise and pride.

We pulled to shore to harvest the fillets while the others found and made camp a mile downriver. Paddling into camp, we found tents rising, and the bug tent nearly complete. After unloading, we put the fish in the pan. The oil flamed, and the fillets sizzled. To complement the trout, we baked two pans of golden cornbread, popped corn kernels in oil with brown sugar and salt, and mixed brownies and Oreo cake in our cups. We laughed the entire time about anything and everything, our conversations wandering all around during this five-course meal fit for our band of lost boys.

The next day, we ran thirteen miles of amazing white water, flowing through the river with ease and exuberance, slipping from smooth swift current into the boiling rapids, as frictionless and agile as three drops of mercury.

Mike and I had slipped into a Zen state, and paddling became an absolute joy, the canoe an extension of us, all three pieces working in concert, moving together with no commands but feel and intuition.

Not all the rapids were easy going. Approaching one later in the day, we slowed to scout from the water. We stood up in the boats, gazing downstream. Everything looked within our abilities. Mike and I would be the third boat, and we watched the first as it headed into the rapids. They got a quarter mile down before they became hard to see. While they dipped and turned in the distance, we drifted slowly downriver. It was hard to wait before a set, holding back while an entire river pushed forward, waiting for the adrenaline rush of the rapids, but we held as patiently as we could.

The second boat went, maneuvering around rocks and boulders in the wide, deep set. Mike and I entered the first part, slipping lazily through wide downstream V's. I could just make out the first boat in the distance, dipping and turning. Just as we passed the point of no return, I saw them switch from lazy paddling to suddenly kneeling bolt upright and paddling with conviction. They turned and then dropped out of sight, as if the river had swallowed them.

I stared for a moment, blinked several times, and squinted, looking through the wavering air toward where their canoe had been, trying to make sense of it. It was as if they had flown into an air pocket. The second canoe wasn't far behind them, and when they reached the same spot, they, too, disappeared.

Mike and I looked at each other. We weren't sure what they had hit, and wouldn't know until we got right on top of it. Canoes one and two were gone, out of sight, and we'd seen no signal to paddle. We shouldn't have followed them without a go-ahead. All we knew was that they had hit a feature large enough to swallow them whole, and we were in the thick of the current, heading toward it.

After dodging the first few features, it wasn't long before we were coming up on the drop. The speed and intensity of the rapids was increasing. We'd been wordlessly navigating the easy stuff, but a glimmering line straight ahead of us had grown. It was a pillow, an underwater ridge disguised by water bulging over the top, and we were headed for it.

"You see that's a ledge?" I said to Mike.

"What?" he said.

"We should go left," I said, louder.

"What?" Mike said, louder too.

"There's a ledge in front of us!" I yelled.

Mike finally heard me and saw the line of dropping water, "Oh, shit!" he said.

We were just seconds from it. To either side of the pillow were rocks and spattering, chaotic waves, but just left of it, a narrow downstream V poured between two large features, like water from a pitcher. That stream could be our way through. Dodging more features, we approached the drop, sliding sideways to align with our chute. I pried to push the bow over the last foot, and we caught in the smooth arc of water. My body tensed, my knees pressing hard against the canoe's sides while my fiberglass paddle dug into a static draw. Just as had happened with the first two boats, we dropped several feet, riding the wave and slamming down with a splash.

With no time to waste, we scrambled for the next feature, just ahead of us. The current flowed much faster as it dropped from feature to feature.

We dug in hard, willing the boat into the next chute. At the last moment, we lined up and dropped. Water splashed over the gunwales, turning the faded nylon of our packs dark blue and green. I shook the water from my face and kept paddling. The river churned all around us. Our strokes, rudders, draws, and pries turned the heavily laden seventeen-foot vessel on a dime. We leaned into turns as if on a racetrack, side-slipping across the boiling stream, shooting down features along a narrow line between boulder and hole, torrent and eddy.

The last feature was not one but two chutes, one behind the other in a hard S curve. We hit the first, then swung sideways, entering the second chute with the boat leaning like a bobsled in a turn. For a moment, we floated there, suspended between sky and raging river. We held the boat firmly with our knees, bodies tense all the way up to our arms and hands and paddles, which vibrated with the water and rock.

After the moment of suspension, we launched out onto the flat water beyond and drifted downstream amid froth and foam from the set. Seeing that we had cleared the last major feature, we whooped with joy. We had been in the set less than a minute.

We passed an enormous boulder and found our compatriots hiding in its eddy. All of them were bailing water from their boats.

Dan glared at us from behind the bilge pump he was working. "You didn't wait for the signal," he said. He was right; none of us had waited. He probably would have had us line the boats down the whole thing from shore. We told him we were sorry, and he asked if we needed to bilge water. Mike and I glanced down to the floor of our canoe. We hadn't taken on any water. By some providence, we'd made it clean through the entire set.

Mike and I were still riding high from the adrenaline as we sat silently on our seats. Watching from upstream, we hadn't thought much of the set and assumed we were good to go, that the second canoe had been given a go signal. Then we had let ourselves drift downstream until we were committed to the run. We should have waited. The river was getting more complicated, and we needed to be more cautious, communicate better, remember to listen, and do things by the book.

With the other two boats bilged, we silently peeled out of the eddy, one by one, and continued downstream.

We paddled for a while before I got out my GPS to check our status. Once it had powered on and connected with the orbiting satellites, it displayed our speed. Nine miles per hour. *Damn.* A normal speed in a canoe, with no wind and without breaking a sweat, is three to four miles per hour—basically a walking pace. Paddling as fast as we could in our overloaded boats on flat water, we would have been happy to push seven miles per hour. We were paddling easily, though, the shore rushing past as if we were riding bikes.

We continued downriver until, rounding a bend, we noticed tall stones that broke the monotonous horizon. *More inukshuks?* No, these were not inukshuks, not piles of stones. Closer up, I could see that these were monoliths of what looked to be granite, rising like Stonehenge from the barren tundra. Curious, we pulled off the river and secured the boats. We scrambled up the steep bank until we were standing at eye level with the stones. They were granite columns, smooth as driftwood. Remnants from the glacier? Maybe, but it had not placed them like this. The stones had been formed and polished by the glacier or perhaps the river, but people had put them here. We were in another ancient site. I thought of the tent rings I'd seen on the island. These were of different stone, in a different configuration for a different purpose, maybe placed by a different people. I remembered travels with my parents since I was three, to countless ancient sites in Mexico, Southeast Asia, and, just a month ago, South America. This wasn't like an Inca complex. This was more … *What's the word? Yes,* nomadic. *Like us.* This wasn't the foundation of a permanent structure. The structures themselves would have been made from hides, wood, and bone and were long gone. These stones helped make the camp, though, and they were permanent—had lasted hundreds, maybe thousands of winters. To one side of the monoliths were four more oblong stones, smaller and narrower than the others, set in pairs eight feet apart—a kayak stand, where the craft could be set safely aboveground. Not far from the stand was a small circle of sooted stones where fires had once burned. Beyond that was a larger circle, at least ten feet wide. *A tent!* The site was scattered with animal bones. We were standing on an Inuit site, a campground from long ago. I looked again at the bones. A hunting camp.

We were surprised to find the site. We had been lulled into feeling that we were the only ones out here on the frontier. It was like thinking you're alone and suddenly realizing you're not. I had pictured the tundra as a barren wasteland, and it so often felt as if we were its first explorers. That's more a testament to the travelers and stewards of the land who came before us than anything else. Stumbling onto this ancient site brought a startling permanence to our surroundings. Those travelers who had made this river and land their home so long ago, who knew it so much better than we did and had spent so much more time here, were dead or gone. These stones had been here for generations since, as unchanging as the tundra.

Below us, the river flowed in a gentle oxbow that wrapped around the site. I looked across it to the far shore and a ridge just like the one where I now stood. I pictured a man's weathered face, hundreds of years ago under a gray sky, hooded eyes looking out across the land, much as I was doing now. I imagined him watching across the Dubawnt as one of his traveling companions climbed up the far shore, to become a dark silhouette against the clouds. The tundra was a place with deep history, and only now was I truly struck with the reality that we were traveling in someone's homeland. I felt the presence of all those who had come before us, and how temporary we were. Someone had tended a fire here, cleaned these bones, sheltered against the cold.

We returned to the boats quietly, avoiding any disturbance of the site. I looked back and imagined those tents of skins, the graceful curves of a taut kayak, and that weathered face behind a ringed hood of furs.

The map showed that we were approaching a violent drop of eight feet, with a portage Dan had read about in the journals. We pulled off the river and carried our gear, looking in wonder at the beautiful waterfall—the energy of the river made visible. It pounded over the ridge, the sparkling cataract glowing gold in the sunlight. We dropped our gear at the far end of the portage and decided to make camp early, by the falls.

The pools downstream from the falls were filled with lake trout, and Darin even caught a grayling, our first of the trip. He held it up with quiet pride. It had a large sailfin, and a coat of dark scales that shone like a slick of oil. The rich, delicate white flesh was excellent.

The sun and moon were the same size that evening—huge, glowing the same dull orange from opposite horizons as day ended and night began.

When we were taking down the tents the next morning, Dan pointed out two dark shapes moving across the horizon. We squinted across the rolling grasses. Was this a pair of grizzlies? As they lumbered along, we could make out a curving helmet of heavy horns and a stiff, matted skirt of fur. It was two musk oxen. We set down our emptied cups of rice and raisins and ran toward a low ridge to see them more closely. The closer we got, the lower we stooped to keep below their line of sight, until we were crawling our way to the crest. Peering over the rise, we sat for several minutes until the beasts had closed to within a hundred yards of us. The naturally ill-tempered creatures finally took note and looked at us with a defiant, annoyed stare. Their dark brown eyes were set, their long outer hair swaying in the wind.

"Okay, everybody," Dan whispered, his eyes locked on the musk oxen, "back up slowly."

We began backpedaling and stood up slowly. The musk oxen continued staring and now turned to face us. Darin hadn't moved.

"I'm serious, guys. Let's get out of here," Dan said.

Darin seemed entranced. Dan yelled, and this time Darin relented, slinking back from the crest. We backed cautiously, eyes locked on the sullen gaze of the prehistoric-looking beasts. When we had reached the campsite, the musk oxen seemed satisfied and went back to their aimless plodding. They seemed wary, though, turning to watch as we packed camp and loaded the canoes. *We* were the intruders into *their* home.

It wasn't until we were on the water that we could relax. It had been a fascinating encounter, but it left me uneasy.

> The caribou feeds the wolf, but it is the wolf who keeps the caribou strong.
>
> **—Inuit proverb**

CHAPTER EIGHT

Leaving the Dubawnt River / Trailblazing: Days 24–25

The next two days were long, filled with white water and difficult navigation. We all ran into more rocks than we should have, the worst incident being when Auggie and Jean broadsided a rock and tipped enough that water spilled over the gunwales. They righted quickly and nothing came of it, but all of us were sloppy. We were approaching the end of our time on the Dubawnt, and I felt I no longer knew the river. It had changed.

We paddled into a labyrinth of densely packed islands and narrow peninsulas. Mike, reading the map in another canoe, struggled to decipher the choking maze of islands. I was disoriented. It felt like a dark fairy-tale swamp, the air strangely thick. Arctic terns dive-bombed us, as if telling us to turn back. White and black feathers flicked violently in the air around the boats, their sharp squawks piercing the heavy silence. As we approached the far edge of the little riverine archipelago, the water narrowed, islands on either side melded into the shoreline, and we slipped almost imperceptibly into the vacuum of the river.

It wound tightly for miles, dropping into a deep channel. It tumbled into white water several times before we stopped for lunch at a huge eddy with sand beaches, snowbanks, and a steep esker. The river here was hidden in a sort of canyon formed by high eskers on either side. We climbed one to see where we were.

At the top of the esker, we consumed our meager lunch calories while mosquitoes flew in swarms around us. In the strong wind, they hung behind us like shadows, buzzing in the lee created by our bodies.

Reviewing the map, we discovered that this esker marked the end of our time on the Dubawnt River. A quarter mile downstream, we were to make a portage over the esker, onto a nameless lake that I imagined few people had ever paddled on. The Dubawnt, rushing by us along the bottom of the esker's steep north side, churned at the silt before quickly curving back away to the northeast. This was the shortest crossing. To the south was new territory.

This esker was the point where the unknown part of our trip truly began. Unlike on other legs of the journey, here we had no journals telling us what worked or what to expect on any of these rivers and lakes. All we had were our maps and compass. In that musty map room at the library, Dan had pored over charts and found a traverse that would take us from the Dubawnt over unnamed lakes and across enormous eskers before dropping us onto Tebesjuak Lake and the Kunwak River. The Kunwak was known to Menogyn and common enough in the journals, and the Femmes were now paddling it. For the next two days, though, until we reached Tebesjuak Lake, we would be bushwhacking.

Before we packed up for the portage, we stood for a group picture on the esker, both parts of our trip in the background: the Dubawnt on our right, and the new unknown lake on our left.

After a hard triple-back on the short, nearly vertical portage, we were loading our boats on a perfectly still lake with no name. We plotted a course alongshore and dipped our paddles into the water, the ripples from our three boats moving silently across the placid surface.

My mind was elsewhere. I had understood our route and some of its eccentricities and challenges before we stepped off the floatplane on Wholdaia Lake. I had imagined a solid line, squiggling and winding its way northeast into Nunavut, traveling from a land of low trees and big lakes to barren tundra. Now that imaginary line was gone. It had stopped at this esker. I could no longer imagine what lay ahead, and it worried me.

From our boats, we eyed a part of the esker that jumped up to a peak. We pulled off the lake, pulling the boats up to a solid beaching before hiking up to make camp. Above the steep slope, the ground flattened but for a deep crater to one side and an abrupt hill far behind it that was so covered in rocks it looked like a hedgehog. I found dry grass and hard-packed dirt that would make for good tent staking. The wind had begun to blow, and we rolled out the tents, setting them up and staking them down as quickly as we could, guying them out at every possible point.

With both tents pitched and secured, I paused and looked to the north. In the distance, I could still see the Dubawnt and the canyon it carved, wending and bending through the empty flats. When the wind ebbed, the rush of its waters wafted over, the sound flickering like the white noise of a far-off highway. It seemed to be calling out, reminding me not to forget it, or telling me that it was still there, just in the shadows beyond the esker's edge.

All the while, Dan had been making a stew in the big pot, with chopped vegetables, stock, and butter. It was nearly done now, and a heavy loaf of bread was baking in the outback oven. It was delicious. We all ate more than our fill.

With dinner done, we struck the bug tent and laid its limp mass on the ground. We weighted it down so it wouldn't blow away. In a post-binge stupor, we grabbed a single Nalgene, an empty gas can, an empty tent bag, a pair of boots, and an antler. After all these items were on the bug tent, we looked at it for a long time. In our foggy malaise, we had grabbed all lightweight things to weigh it down, and some things that we would still need before the end of the night. The six of us looked at one another, aware that we had worked to accomplish exactly nothing and suddenly realizing how tired we all were. A contagion of giddiness spread almost instantly. Scowls quickly turned to furrowed brows and barely contained laughter. Pursed lips parted into smiles. My lips, tight against each other, couldn't hold back the chortle behind them. Almost in unison, we all started to laugh. Our giggles rolled like the hills around us. We were in such a mood, it seemed even the wind could kick up a gale of laughter.

Overfull from dinner and tuckered out from all the fun, I slipped into the tent. Though I could still hear the hum and rush, we were no longer on the Dubawnt River. The Kunwak and the Kazan would be our next chapters.

I rolled onto my sleeping bag and got out my journal, balancing it on my makeshift pillow. On the stiff brown paper, I penned everything I could remember about the day. In my scratchy handwriting, I concluded the entry:

> *Today we left something that's been with us for almost the entire past month. Now we are on a new section of the trip, and a whole new water system. Tomorrow, we will be doing a four-mile portage from a no-name lake the size of a Boundary Waters Canoe Area Wilderness lake, and just twenty rods from the Dubawnt, to the Kunwak River. Once on the Kunwak, we should be two days behind the Femmes. With our trailblazing today, a small and mysterious strip of land in the back of Dan's mind for the past six months, a leap of faith based on contour lines on a map, was actually seen and conquered. We can only hope that tomorrow's endeavors at "firsts" come with equal success.*

I replaced the cap on my pen and stowed it and my journal in a dry bag just beyond my pillow. Rolling over, I felt the creak of my tired, sore body, especially my abs after all the laughter. Pulling up my sleeping bag, I let out a relaxed sigh and closed my eyes for the night.

Moving in the morning was difficult, but we forced ourselves out of a deep sleep and wriggled free from our sleeping bags.

On the nameless lake, we paddled alongshore, following the lazy arc of the esker to what we estimated would be a four-mile portage. Halfway across the lake, we saw a pair of musk oxen lumbering across the open ground. They ignored us as we paddled quietly toward them to watch and take photos. Dan put his elbows on his knees and peered through binoculars. I zoomed in with my camera lens. Despite the animals' slow plod, they quickly reached and passed us.

With that distraction now out of sight over the crest of the esker, we set a course for the start of our bushwhacking portage. We pulled off and began unloading in a green, bush-choked swamp full of low vegetation and grass tussocks. From this, the esker rose like a siege wall. Reaching it, we angled up the steep stone to cut a few hundred extra feet from our trip. Sweat beaded on my brow, and I breathed heavily. It was a long climb to the top. We plodded up quietly, no one talking, the packs pulling heavily at our shoulders. The constant slope didn't afford any good ground for breaks, so we pushed on to the crest.

The top of the esker was like a gravel trail. It was about the nicest walking surface you could ever expect in the backcountry, like walking on an old driveway. I gazed at the winding crest as it meandered across the tundra, swooping with graceful curves like the river that had birthed it, bending left and right at the whim of ancient forces.

It was a long haul with a heavy load, and at the end I was only too happy to set my Pelican case and camera down before heaving the pack to the ground. We groaned and stretched, pleased with our achievement but still with so far to go. We were not even halfway done with the portage, and after too short a break, we turned and quietly marched back for the other load.

Almost halfway back to the start, I eyed two small brown objects on the ground. Their uniform color and shape seemed out of place in the tundra. I stepped closer and found parallel lines and straight edges. They didn't reflect much light, and then I realized that they were old rifle cartridge casings. I turned the two tarnished shells in my hand, looking at the dimple from the firing pin and reading the stamped inscription around it: WIN 30.30. They looked old, the color of deep bronze, burnished to a subtle sheen over years of dust, sand, and snow.

Turning them in my hand, I tried to divine their history. I imagined the dark iron sights of a lever-action rifle. A caribou grazes in the flats below. The barrel tracks a moment before the hammer drops. The explosion echoes, alerting the caribou a moment too late. That first shot misses, and the caribou runs. The hunter racks the lever, launching the spent casing with a twisting trail of white smoke and pushing another round into the chamber. The second shot cracks like close thunder. He quickly ejects the second shell, chambering a third. But the caribou has fallen. The hunter, clothed in brown fur, lowers his rifle and quietly walks to his kill. He utters the beginnings of a prayer to the animal's spirit, leaving the two warm casings to cool on the hard ground.

Finding the shells served as a reminder that even though we were bushwhacking, we were not the first people to walk this esker. We were guests here. The tundra of Nunavut had been home to the people who made that stone kayak stand, the man who fired those shells, and the countless others over the centuries. This ground held a history we barely understood, and secrets we would never know.

I suddenly felt a power in this place, something much bigger than I and more sweeping than our transient presence here.

I closed my hand around the shells and dropped them into my pocket. We turned and continued our march back to the beginning of the portage.

Hours later, as the sun was setting, we dropped the last of our gear at the end and celebrated in the bug tent, eating a delicious green pesto pasta and gobs of hard candy from our hoard. We were exhausted. The portage had taken nearly half the day, but we had reached the shore of Tebesjuak Lake. If everything went as planned, we wouldn't have another portage until Kazan Falls, ten days later. But that was after Princess Mary Lake. It was a future I couldn't imagine.

> Now is no time to think of what you do not have. Think of what you can do with that there is.
>
> —**Ernest Hemingway,** *The Old Man and the Sea*

CHAPTER NINE

The Kunwak: Days 26–27

I felt consciousness arrive like the dull flicker of a fluorescent light powering on. Blinking at the tent's nylon ceiling, for a moment I didn't know where I was. When I tried to summon my memories from the night before, they wouldn't come. A general realization formed as the memory of the arduous portage rolled through my mind like a slow fog. As I remembered our general location, my mind began to rotate, orienting itself to where I now was. Finally, things began clicking into place, though I was still unsure where our tents were pitched. I emerged from my sleeping bag and got ready for the day. Soon, I was dressed in grubby trail clothes, had my bag stuffed and my pad folded, and was ready to go. As I unzipped the door of the Prophet, the pale khaki gravel of the esker appeared. I looked down onto a distant Tebesjuak Lake, a slight breeze turning it a shimmering blue against the drab stones. With the door open, my orientation adjusted, meshing with my mental map of Nunavut. The puzzle was complete.

The sun, already high above the horizon, was shining brightly through a cloudless blue sky. After breakfast, we walked to a sandy point to rinse off with a quick swim. The water was barely warmer than when we swam among the icebergs in Dubawnt Lake. I came out of the water and stood dripping in the radiant sun, relishing the warmth as it spread across my back and neck, down to my fingers and toes.

As we pushed off from camp, we were preoccupied. Though we were completing our bushwhack section of the trip, we might not be alone any longer.

Normally, we wouldn't have been thinking about running into other people, but today was different. Since the beginning of July, we'd been meandering our way through the Northwest Territories and Nunavut on our own path. At the same time, the Femmes had been on their own winding route to the east, on the Kunwak. For the weeks since our trips started, there had been no possibility of seeing them. Today, our routes would merge. It was unlikely that we would run into them, since their itinerary put them several days ahead of where we were. Still, we scanned the distance and kept our voices low.

By lunchtime, the air had stilled considerably, leaving the water calm and flat like glass—a brilliant reflection of the pale-blue sky. We skimmed quietly through, suspended between sky and water.

A mile from where we planned on camping, I prepared a rod for trolling. We still had time to catch fish for dinner, pull over, and fillet it before we stopped for the night.

"What do you want?" Darin asked, rifling through the tackle box.

"Give me a five of diamonds," I said. He searched, and found the large yellow Dardevle spoon, its five red diamonds bright in the sun. He snapped it onto the steel leader and draped it over the side of the boat.

"There ya go," Darin said.

We started forward, and I cast into the still water. I let out a lot of line, dropping the lure into the cold depths before closing the bail and propping the rod behind my leg. Satisfied with the rig, I picked up my paddle, and we started trolling.

Soon, the pole bent hard. I reflexively clamped on to the rod. It curved 180 degrees. I had experienced this kind of sudden force before when snagging an immovable object, and once when an enormous walleye latched on and just sat there like a stone. With the walleye, I'd had delicate ten-pound-test line and had to be careful not to break it. This pole had twenty-pound-test line. This would not be a game of finesse, but rather a tug-of-war that would have made Hemingway proud—provided it was, in fact, a fish.

Then I felt a tug. This was no snag. Suddenly, it took off. The drag whined as line spun out into the lake. I twisted the knob to try to lock it down and cranked on the reel.

All three boats surrounded my line, and our boat spun around in three full circles at the fish's whim. With line still spinning out, I ditched the reel and hauled the fish toward the canoe hand over hand, its fight intensifying the closer it came to the light. Finally, it burst through the surface in a great, splashing flash of silver. The fish flailed within the tight triangle of boats, slamming our hulls with heavy thuds. Water flew in all directions from flapping fins and paddles.

We all tried to land the beast. Finally, Darin leaned as far out of the boat as he could, latched on to its tail, as he had with his own fish days earlier, and pulled it out of the water and into the boat. He somehow managed to hand it to me, and I suddenly appreciated its size. It was the length of my leg and bigger around than my thigh. I worried that it might convulse at any moment. Mike took a photo, and I carefully lowered the behemoth into the boat.

The fish filled the entire bow of the canoe, its head against the hull at the front, the tail grazing the forward edge of my seat. When it thrashed, the muscular tail came within an inch of my groin. With my feet on the gunwales, I tried to keep out of the way, and we paddled toward shore as fast as we could.

The fillets were huge, inches thick, like two long, bright steaks. They nearly filled our big pot. As we paddled downstream toward camp with our catch nestled between the portage pads, the sun burst through far-off storm clouds, bathing the tundra in light. Everything turned to patterns of sunset, colorful sky all around us.

By the time we reached camp, all the tents were set up. We entered the bug tent and presented the fillets. Auggie worked the skillet, cooking the monster lake trout, tossing it in the pan like a true chef. It was mesmerizing—and delicious.

We cooked, ate, and laughed, finally making our way to the tents for the night. After we had shut the doors and zipped our bags, we realized we'd forgotten to get bear spray for the night. I interrupted my journaling and stepped out into the buggy cold. I padded to the beached canoes and rifled through the packs. I found the bear sprays, each tied shut in one of the poly bags like those we used for food. I imagined dutifully untying it in an actual bear encounter.

"So are we supposed to untie this if we ever need to use the bear spray?" I asked Dan through the tent wall.

"No," he said. "If you ever have to use it, you just rip the plastic."

"Oh," I said, turning the can in my hand. "Right."

Morning came gray with a breeze that stirred the water into cloudy steel. As we collapsed the tents and stuffed our packs, the wind stiffened and the water darkened. While we ate breakfast, waves began to flash with whitecaps. We cleared camp quickly, knowing we were in for a fight.

After an hour of hard paddling into a stiff headwind, we felt the familiar pull of current again as we slid into the Kunwak, the second major river system of the trip. We followed it around a corner and came upon what looked like a musk ox on the far shore. Curious, we continued toward it. As we neared it, we realized it wasn't a musk ox at all. There were two bright-red boats, surrounded by five figures.

Startled, I stopped paddling and looked back over my shoulder to Dan. He continued paddling. "We're gonna stop, right?" I said.

"Well," he said, squinting toward them, "I think it'd be weird just to paddle on by."

The Femmes had stopped what they were doing and were watching us. We turned toward them, and I was suddenly nervous about seeing friends, people I knew outside our group.

In a moment, we had reached shore. Skidding onto the gravel, we climbed out and hugged one another. So much had happened since we last spoke, but it was as if we'd just said goodbye at the floatplane base. Conversations picked up where they'd left off, and we shared stories of our adventures.

"You've caught *fish*?" the girls exclaimed.

"Yeah, lots of 'em!" we said. They looked at one another, brows furrowed in disbelief, then laughed as if they had known they were using their tackle wrong.

We told them how short we were on food, and they told us how much extra they had. We learned that they were sleeping normal hours, and they were surprised at our sometimes twelve-plus hours of sleep. We were on totally different trips, both equally intense, each group loving their own journey.

Our talk wandered comfortably but eventually turned to business. We would make trades. We would catch and fillet a fish for them, and for this, the girls would give us several meals' worth of pasta, as well as lots of chocolate, fish batter, and a collection of spices, including cinnamon, which we'd nearly run out of. We would exchange the goods the next time our groups met.

We hugged again and said, "See you later!" Wanting to give each group space, we stayed on shore for an early lunch while the Femmes gathered their things and pushed off into the Kunwak, turning downstream, into the wind.

By the time we were back on the water, the wind had kicked up stronger still. The river wound in tight curves, the water clear, deep, and swift. Held back by the wind, we could barely feel the current. It played with our boat, and the wrong paddle stroke made the canoe spin unnaturally with the contrary forces of wind, waves, flow, and up-currents.

After hours of fighting our way downstream, we rounded a corner and found the Femmes in a wide eddy, under an embankment that shielded them from the wind. They had held some of their goods from lunch for trade. “We’ve got *dark* chocolate, so we don’t even want these,” said their guide, handing a bag of Hershey’s bars over from her canoe. They passed bags of fish batter, too. We were thrilled. We would give them their fish once we caught one on the next lake.

Thanking them, we turned downstream. Shore slipped past, and we accelerated in a deep, protected channel. Before long, we were spat out into a lake, again whipped by wind blowing unhindered in the vastness of tundra and open water. Gradually, it eased into a steady breeze, and the rough waves calmed to a gentle lapping against the canoes. We relaxed enough to throw a lure out and troll.

Before long, Mike yelled, “Fish on!” and we headed for shore with our catch. We would be able to keep up our end of the trade right away.

Mike and I worked on the trout while Darin and Auggie filleted another they had caught. Soon, the Femmes had their fish and were pushing off to make camp downriver. We found our own spot about a mile down, off a sandy beach that afforded a good view of the sunset. We added the trout to a rich macaroni and cheese. The pink cubes of poached fish were a deliciously juicy addition. As the food cooked, I napped, exhausted, in the bug tent, a Nalgene as my pillow, watching the sunset between dreams. Things were going well.

I am incapable of conceiving infinity, and yet I do not accept finity. I want this adventure that is the context of my life to go on without end.

—**Simone de Beauvoir,** *The Coming of Age (La Vieillesse)*

CHAPTER TEN

Fate: Day 28

The next morning, we made our way to the Femmes' camp for final swaps. We traded the filleted fish, a broken pair of binoculars, and some fresh sprouts we'd grown, for a cache of cinnamon, chocolate, powdered milk, soup, shore lunch, mixed veggies, dried fruit, dried red and green peppers, refried beans, and spaghetti. Despite the apparent disparity in our favor, both groups were happy with the trade. Excited, they took their goods, and we took our food. With this bump in our stores, we no longer had to worry about running low on food. We'd needed the staples, but I was most looking forward to the cinnamon. We were nearly out and planned on making cinnamon rolls the next morning.

We said goodbye to the Femmes for what would likely be our last close encounter of the trip. If everything went smoothly when they passed us the next day, on our layover day on Princess Mary Lake, we wouldn't see them until the end of the trip, close to Baker Lake.

Paddling away from their site, we made a long seven-mile crossing on Mallery Lake and slipped back onto the Kunwak. The surface was smooth, like dark blown glass, and every few moments, little jewels of river stones glimmered from the deep.

Only a few turns into the river, we encountered something utterly foreign and unexpected. In the distance was an arrangement of straight lines and decisive angles that gradually defined itself into four crude walls and a roof. It was a shack. Only ten feet tall at most, it loomed like a mighty tower above the treeless expanse.

Why was this building here? What compelled someone to haul materials to this place and build a shelter?

Of course, we had to explore it. We made our way across the tundra to the decaying hovel. It had peeling wooden walls with debris all around: rusted tins, husks of old tools and machinery, and scattered animal bones. The closer we got to it, the thicker the clouds of blackflies and gnats became. I felt them hitting my pants and slapping against my jacket. They weren't going after us at first, as if they had run into us without realizing we were there. Almost as soon as we reached the hut, we all started to inhale bugs from the humming cloud. It took us only a few moments of coughing and sputtering to overcome our curiosity of the unknown. I couldn't leave without looking inside, though. Squinting and holding my breath, I peeked through a small pane of weather-hazed glass. Inside the hut were the dark-gray outlines of a time long gone: food cans, glass jars, cups and plates on crooked shelves nailed to the wall, a narrow countertop of bloated particle board, several pots on a crude stove, its rusted flue wandering up to the peaked roof, with gray splits of old wood nearby on the dirty plywood floor. Barely able to breathe in the thick cloud of bugs, I pulled back from the window and we retreated. We were still coughing and spitting when we reached the boats.

"Let's get out of here," Dan said, shoving off from shore.

In swift rapids, we maneuvered quickly and deftly around boulders like giant eggs. Sliding through downstream V's and dodging the smaller rocks, we slipped gently onto the placid surface of a delta that widened beyond my peripheral vision. Princess Mary Lake.

The lake was a hidden gem, a view that could only be earned. Spread out before us was pristine water twenty-five miles across, surrounded by a shoreline of steep, rocky hills. As I followed the bending horizon, a shimmering image arrested my gaze. This was not something we had expected to encounter again. It looked like outstretched ice floes. Concerned, we turned our boats abruptly and pulled off the water to get a look from higher up.

We landed on the north shore and began a long trek up one of Princess Mary's mountains. The hike up was deceptively long, the gentle slope punctuated with steep grades. The view from the top looked over much of the west end of the lake. We shielded our eyes from the bright sun. There was nothing but sparkling water and dark knurled tundra as far as the bend of the earth would let us look. We were relieved. The mass of ice we'd seen had been a mirage.

With that out of our minds, we marveled at the beauty around us. In the distance were the peaks of mountains even larger than the one we now stood on. In the middle of the lake, a large crescent-moon island towered above the surrounding peaks. It stood vigil over the entire lake and the tundra beyond, shining in the late-afternoon sun. Looking at it from across the miles of glassy water, we decided to make camp on it. Somewhere on that promontory of crumbling talus, scree, and tundra, we would find a hospitable flat spot, then spend the next layover day exploring.

When we got back on the water, the Femmes were paddling in the bay across from us. Waiting to let them pass, we gorged on trail mix and candy. Soon, the Femmes stopped and made camp, their figures bustling about on the distant tundra as we paddled toward our island.

As we slid over the mirrored lake, the sun dipped slowly in the distance behind us, bathing everything in a muted rainbow of color. We paddled our three boats in a tight formation and talked quietly back and forth. Our conversation wandered to randomness and the predictability of chance. Dan told us about chaos theory. "If every single atom were to be mapped at one instant, every particle known at one given time," he said, "the interactions between those atoms could be calculated and the future could be predicted."

Dan dipped his paddle and went on. If everything and every happening is calculable, then, theoretically, all is preordained. None of us was completely satisfied with the theory, and we quietly debated between our three boats. How could a system based on chemical and electrical reactions have randomness? What is random, and can anything be truly random? Every instance and happening stems from the interconnectedness of everything on the planet and beyond. An asteroid struck from a distant world floats through space, colliding with more and more pieces, drifting along as if in a kaleidoscope of interactions, until it crashes into the moon. Its destiny was written when it started its journey, or even before that.

The idea of predicting the future led to a discussion of fate and how this scientific version of future prediction was essentially a "quantifiable fate." Fate could be spiritual, or it could be chemical and physical, if it exists at all. I didn't believe in fate, but I believed in purpose and that some things didn't happen by chance. I believed that things happen for a reason and that we may or may not ever understand the why, the how, the results of what is happening, or our influence on it.

Left to our thoughts, we paddled on in silence. Our boats slowly drifted apart until each was alone on the vast lake. *I believe in free will,* I thought, *though we don't always have desirable choices, or make the best choice from those available.* But free will didn't seem to have a place in that calculable system. I didn't want to accept this. It didn't add up. Things happen to us, but we are the unknown variable. Factors and forces confront us, but we are not simply circumstances in an algorithm—we choose and decide and act. That is not definite. The human element is an unknown quantity. *And yet,* I thought, *things somehow do happen for a reason.*

The sun touched the horizon and then sank below it, and I felt a chill. I continued pulling toward the island in a state of reflection, thinking about my own life. As we approached it, the island loomed larger and the shore grew steeper. When we finally arrived, I was amazed at its size. We came to a stone-and-grass landing so gradual that it seemed to extend the plane of the lake. Then the ground quickly angled up and kept rising, climbing in rolling waves for several hundred feet before jutting sharply up into a cliff.

It was somewhere around midnight when we began scouting for the site. The steep slopes of the island would have made a terrible campsite but for a series of plateau steps far up from shore. A third of the way up the hill and several hundred yards from the water, we pitched our tents. Behind our site, a steep cliff towered high above us in crumbling rocky terraces and stone escarpments. This was one of the best sites of the trip! From camp, we had a commanding view of the tundra, half of Princess Mary Lake, and the steep, stony mountains and rolling hills beyond. We wondered what more a view from the top would hold, what we would encounter on the ridge. It had to be breathtaking. We would find out tomorrow.

As the hour passed into the next day, July 31, our twenty-ninth day on trail, the sun started to brighten the eastern sky, and we made our way to the tents, to bed. Sometime the next morning, the Femmes would pass us for the last time before the end of the trip, most likely while we were still deep asleep for the start of our layover day.

In the tent, tucked halfway into my sleeping bag, I got out my journal and unsnapped the leather cover. I flipped through page after page of tiny inscrutable handwriting until I reached blank space. I wrote about the day, the food swap, and the lake, about the sunset and paddling and the island, about fate and free will.

It's amazing to think that we have around ten days left of the trip. To think of the change that's occurred to me and the rest of us is also amazing. It's one of those things that you know has happened but there are few signs of it apparent to you, and we won't know how we've changed until we go back to our normal home life. Right now the way we are and whatever we've become is our normal day-to-day life, and those changes have been seen from an inside view over time. Once our perspective goes back to the one we live in most of the time, we will see how we've changed, as well as have a new outlook on life overall. All these things will truly affect the rest of our lives.

I capped my pen, shut my journal, and put them away before turning over to my back and zipping my sleeping bag shut.

It would be a week before I opened my journal again.

PART II

Life is a storm, my young friend. You will bask in the sunlight one moment, be shattered on the rocks the next. What makes you a man is what you do when that storm comes.

—**Alexandre Dumas,** *The Count of Monte Cristo*

CHAPTER ELEVEN

A Layover Day: Day 29, 10:00

When we all woke, we prepared camp for the day, putting away things from the night before and helping Dan with his backcountry baking.

Before long, a pair of canoes appeared on the lake. The Femmes were paddling from their camp across the lake, southeast to the end of the lake, then downriver. As their tiny silhouettes were about to paddle by, they abruptly changed course and turned straight for us. They landed on shore near our boats, yelling up to us as they got out of the canoes and started the long climb toward our camp. Their words were unintelligible.

As they got closer, we started to understand them. They flailed their arms like great, clumsy herons, yelling, "Danger!" "Impending doom!" "Bears!" When they finally reached our camp, they explained that they'd just seen bears on the island, around the point from our campsite. They described a whole group of them, lazily grazing. This didn't sound a lot like bears, which don't usually hang around in a herd together. We asked more questions, and they described the bruins as large, dark-haired creatures. In addition to moving about in a herd, they were tall, always stayed on all fours, and had long hair and long, spindly legs. What the Femmes had seen were surely musk oxen. They had seen neither bear nor musk oxen on their trip until just now, and these were a long way off, so the mix-up was understandable.

We thanked them for their warning, and again for the cinnamon that Dan was at that very moment baking into our sweet rolls. We said goodbye, knowing now that we wouldn't see them again until our journey's end at Baker Lake. I watched as they sauntered back down the hill in their knee-high neoprene boots and brightly colored bandannas and life jackets. Shortly, they were back to their boats and on the lake, shrinking into the distance. I watched as they moved their paddles in the calm waters of Princess Mary Lake until they disappeared behind the edge of the island's steep slope.

As breakfast baked, we lay in the bug tent, each of us deciding how to spend our hard-earned day of free time. Dan talked to us for a bit about the musk oxen, but we weren't too concerned with them. We'd seen them before and knew to be wary of them.

The day was bright, with deep blue shining between wispy clouds that drifted high above. The wind had slowly been picking up, and by then it was a constant breeze.

When Dan told us the cinnamon rolls were done, we converged around the stove. A cloud of steam rolled out of the heat shield when he opened it to reveal the gooey rolls bubbling below. They looked delicious. We waited impatiently while he sopped our portions out of the pan, balancing them precariously in our overfilled cups. The steaming rolls seemed to glow in the sun. My mouth watered as I appraised their shimmering, sugary spirals and the spiced sauce that overflowed from every crevice.

The dough was that perfect mixture of soft doneness and delectable rawness that is difficult to achieve intentionally but almost always happens on trail. Relishing the texture, I carved a piece out with my spoon and blew on the hot morsel until it was safe to eat. When my taste buds hit the sugary buttered sauce, my bliss suddenly changed to surprise, as if I were sipping milk that turned out to be orange juice. Something was wrong. This was not the flavor I expected. The rolls had been tainted, their sweetness gone. My brain registered a bitter dryness. Almost in unison, we looked to one another, confused, our mouths open with half-chewed bites. There was no cinnamon, no nutmeg, no taste of Christmas. It was wrong.

Dismayed, we eyed the rolls with suspicion, then ran to examine the plastic bag of spices. It was the bag that the girls had given us in our trade. This wasn't cinnamon; it was allspice! It seemed the worst thing that could happen on such a perfect day. They had given it to us in good faith, we mused, innocently confusing the two brown powders. Even so, I wanted to tell them, explain the difference, as if it would somehow change things. But the girls were long gone. We had no recourse. The cinnamon rolls weren't cinnamon rolls at all, and that was that. Despite our disgust, on trail you only have so much food, and even if you hate it, you eat it. Even if it's allspice rolls. We slowly chewed the rolls, washing them down with big gulps of water.

Luckily, the best thing about finishing breakfast is that you get to start thinking about lunch. We were only too happy to make the transition. After forcing down the last of the bitter and unsatisfying breakfast, none of us felt sated, so we immediately began preparing a lunch of quesadillas. We lazed around the bug tent, prepping ingredients, frying the tortillas, reading and talking. When the quesadillas were done, I ate my lunch and was immediately sleepy.

The rest of the group prepared to climb the ridge above our campsite and explore the top. But as they gathered their books and cameras, I settled in for a brief nap. I would lie down for a bit, maybe half an hour, and then join them at the top. I would be the only one in camp, but I just didn't have the energy to start up the ridge before having a rest. As they got the last of their gear for the climb, I slid into my sleeping bag, feeling cozy in my long underwear. I pulled the lofty baffles of the sleeping bag up around my chin and relaxed instantly. Wrapped up, I looked out the door of the tent and saw Dan waiting as Jean and Auggie, the last of the exploration party, grabbed their things and started up the ridge. The group took long, efficient steps up the steep slope, as if climbing enormous stairs. Each of them gazed up at the ridge. They disappeared past the seam of the door, and I heard their voices drift away over the wind as they made their way up. I was alone and comfortable. With their voices gone, I was left with nothing but the slow steadiness of my breath, and the gentle sway of the wind as it came and went. Under my head was a carefully folded pillow of shirts and fleeces, custom formed to my head. I closed my eyes. The brightness of the day quickly faded to black.

What should have been a half-hour nap became much longer, though I'm not sure how long. What I do know is that I was in the middle of a swirling dream when I was jolted awake. It felt like being violently ripped from a painting, breaking through underlying layers of color, and ripping through the bright canvas into a gray reality. I was falling backward, then forcibly thrust to the surface, gasping for air.

As if pushed, I jolted upright in my sleeping bag. My eyes popped open, and I gasped. I'd overslept, and I was suddenly struck with the feeling that I was late for something.

I had to climb the ridge—now.

If you are afraid, change your way. **—Inuit proverb**

CHAPTER TWELVE

A Chance Encounter: Day 29, 19:20

The uneasy feeling of being late for something, though I didn't even know what, propelled me fully awake. I needed to get to the top of the ridge as soon as possible. Through the mesh of the tent window, I could see the bug tent, and Mike and Auggie stepping back into it. Darin and Jean were only a few paces away. They were all back from the ridge except Dan. I was late.

I slid out of my sleeping bag. Over my long underwear, I threw on my quick-dry shirt and pants, fleece jacket, and wool hat. Clambering out of the tent, I strapped on my sandals and grabbed my book and the hard-shell Pelican case with my camera inside. I was still out of breath, still waking up, pushed by some unknown sense of urgency.

With long, purposeful strides, I angled toward the foot of the slope.

Dan had just started down from the ridge when I started the long climb up. The hill was steep, a mix of dirt, grass, low shrubs, and loose gravel forming uneven steps that were sometimes taller than I. I had to negotiate a route and scramble my way up. Watching my unsure progress, Mike yelled through the curtain of the bug tent, "If you die, I'm going to eat you!"

"Right," I said.

At the midpoint, I met Dan. We paused on a narrow landing to talk for a moment.

"Yeah, you're about halfway," he said, his chin forward, eyes squinting against the bright sky. "It levels out and then kind of over that way," he pointed through the hill, past the crest of the ridge, "there's a cairn I put back together that had fallen apart." He described the martian landscape at the top, the openness, the emptiness. He told me that while they were up there everyone had just sat and read and written and drawn, but he was happy I had my camera to photograph the space. "From up there, you can see everywhere." With a quick update on dinner, we parted ways.

I set off in earnest to finish the climb. The awkward brick of the camera in my hand threw my balance off, and I often traversed the wider ledges to find shorter steps that I could easily climb using only one hand. The second half of the ridge was easier but steeper. I must have been getting used to the climb. Below me, the slope dropped precipitously down to the field of tents. They looked small now.

As I reached the top, the steep, stepped grade eased to a relatively flat plateau of glacially scarred pink-and-gray granite domes. Among the rolling domes were the occasional tufts of tundra scrub grasses and bushes, but the ground cover was sparse. All around were small boulders and bits of broken, jagged stone ranging from small gravel to large man-size pieces. Cut from the same gritty granite, the stone went on for what seemed like miles, rolling along the sweeping arc of the island to its far shore nearly five miles away. The scene was lunar, empty, rugged.

This was near the highest point of the island, much higher than the peaks surrounding the lake, and the view eclipsed all others we'd had across the tundra. The full 360-degree panorama was already stunning, but I wanted to get to the very highest point, to see *everything*. Even from where I stood, I could see the storms from days ago as they lumbered eastward. To the west, I could see the rain we would be getting for the next couple of days. It was like looking east into the past, west into the future. I could see the shadow of a storm on the horizon, and in thirty-two hours we would have rain. The sky felt close, too. It seemed as if I could stretch my hand out and almost touch the clouds.

Turning my head back down to the granite, I took my bearings and set off in the direction of Dan's cairn.

The granite sand and gravel crunched underfoot as I worked my way up the rolling domes. The wind was much stronger than it had been at the tents, and it blew steadily into the side of my face. I was glad to be wearing my windproof fleece and the layers underneath. I felt the weight of the Pelican case in my hand. The pink of the granite was bright, but every few feet were even brighter tiny tundra flowers that quivered in the wind.

I sighted the cairn and walked toward it with my head down, looking at the details of the ground and watching for boulders. I thought of the summer reading book in my pocket: *The Liars' Club*. I was supposed to read it for my senior year of high school. Come September, we'd be discussing it in my AP literature class. I didn't really want to be reading it on trail, but if you have to do it, it's not so bad on a beautiful crescent-moon island.

I pushed the idea of reading aside and thought of my camera in its heavy, clunky case. Photography seemed a much more enticing activity than reading the book. I'd rather take photos, though I couldn't imagine effectively capturing the feel of the place. I had to try, though. I'd get to the cairn first, drop all my extra gear, and wander around the plateau a bit with the Nikon before settling down for a few pages.

My mind was wandering as I climbed up a large gray dome. Suddenly, something at the crest of the ridge caught my eye. An unexpected shadow. As my synapses fired, the subject of my attention continued moving. There was the flick of brown, still in my periphery. Instantly, before my eyes and brain could take it all in, the moving form set off warning klaxons that flashed images of angry musk oxen. My head snapped upright as my eyes widened, my airway opened, and the adrenal glands began pumping into my bloodstream with a fury I'd never felt before. Everything tightened, my vision sharpened, and my entire trunk burned as my muscles went taut. As my gaze snapped onto the horizon, the brown shape became instantly visible. It was not a musk ox. This was much worse.

Thirty feet in front of me was a grizzly bear. It was on all fours, facing me, snout down. The bear's head mirrored mine, jerking up to attention. It gazed, wide-eyed, clearly startled to find me on this barren plateau. Its head pulled against its shoulders, as if it were trying to step back and stare forward at the same time. Every muscle between my shoulders and hips tensed in alarm and dread, hot like bile. In the milliseconds since realizing the situation, my brain had recalled our bear training from camp. The image of bear mace flashed in one side of my consciousness: how to remove the safety, aim the canister, and prepare to fire—the canister that sat in the tent. In the other half of my mind was Dan's steady voice, saying, "*Hey, bear. Whoa, bear. It's okay, bear.*" Next to this memory was a checklist, growing rapidly like lines of computer code: *Don't run—back away slowly, calmly, speak calmly to the bear—don't run—avoid direct eye contact, keep talking—don't run.*

You're close to helpless, Alex.

FUCK!

"Hey, bear. Whoa, bear. It's okay, bear," I said, not believing it myself as I backed up slowly. It felt as if adrenaline were pouring out of my scalp.

Fuck-fuck-fuck.

"Whoa, bear." My quaking voice was betraying me. I tipped my head down, trying not to make eye contact while still being able to see the bear and whatever it was doing. It stared back at me, not quizzical, not yet aggressive, just wide-eyed in startled momentary indecision.

"It's okay, bear," I said.

With that, it had made its decision. Both bear and boy were in fight-or-flight, and neither was running. The bear launched forward, grunting, in a terrifying false charge, its eyes suddenly set and focused. I carefully backed away in a low, steady stance and continued my mantra. It lurched into another bluff charge. I continued my slow retreat. *Don't run!*

"Whoa, bear. It's okay, bear ..."

My mind whirred through options. I pictured turning and running, sprinting with all I had toward camp. The bear could run almost forty miles per hour. I couldn't. I'd probably stay ahead of it until I reached the top of the ridge, and then what? I'd leap off the edge. My arms and legs would flail in space. Then I'd land hard on splintering bones, in a campsite with five close friends, and an angry grizzly chomping at my heels. *What then?* I'd be broken, defenseless, and they would suddenly be targets, too.

My fight-or-flight response was logical. My brain calculated factors and options faster than I could think, weighing them, discarding the bad ones, then giving me the remaining few to choose from. I picked the only option that didn't guarantee an attack. It was like picking the best of several bad answers on a multiple-choice exam. Every option sucked, but this one seemed to suck the least.

"Whoa, bear," I said.

This next time when it launched onto its front paws, it didn't stop. The infuriated grizzly kept coming. After two more short, powerful steps and bellowed grunts and growls, it stopped. The bear was testing me.

"Easy, bear," I said, shaking.

It launched forward again and charged three steps before pausing. The bear's intensity was orders of magnitude beyond my own hyperreactive state. The bruin stopped for only an instant, a rolling stop, its eyes wide, before leaning into a full charge.

No, I thought. With the increase in violence from the bear, my voice crescendoed. "Whoa, bear, no ... bear." I began yelling hopelessly toward camp, "Guys, bear! Help! Guys! BEAR! HELP! NO! FUCK! BEAR! NO! NO! BEAR!"

In my slow backpedal, I had not increased the thirty feet between us by any noticeable distance. The galloping ferocity coming at me was quickly closing those thirty feet.

As the bear came toward me, I didn't know what to do. I had taken our training as far as it went. There hadn't been a section on what to do when you're alone and six hundred pounds of fury are barreling down on you, other than "*Don't run; stand your ground.*" I stood my ground. When the bear was fifteen feet from me, the ground beneath my feet began to shake from the impact of its pounding paws. My ears began to ring from the grunts and bellows. I stood wide, hands behind me, chest and head leaning toward the onslaught as I let out a loud, guttural all-out roar.

Time didn't slow. My life didn't flash before my eyes. I was struck with clearheaded awareness, accompanied by dramatically decreased reaction time and increased speed, both mentally and physically. My fight-or-flight response was at full throttle, whining at the red line of my tachometer. I was choosing to fight. Unfortunately, so was the bear. We were two bull-headed youths, both terrified—but one was four times the other's size.

The pounding gallop accelerated, shaking the ground. Fifteen feet closed to ten, and like that, the bear was inside that tiny ring of safety. Automatically, I stepped back, pulling my Pelican case behind me. I could feel the bear's growl resonate through my body. It was a tearing, rough sound of rage. My hand reflexively tightened on the Pelican case, gripping the handle even tighter. We had locked eyes. Behind the snarling teeth and the sand-colored mask of fur, the dark-brown eyes bored directly into mine. They were full of intention.

My arm launched forward. The Pelican case swung down, past my shoulder and to the front of my body. As the arc reached the point of release, my hand let go and the fifteen-pound brick sailed off into the cold tundra air. The earth trembled under the creature's hurtling fury. I kept up my steady backward steps. The missile flew through the air, tumbling slowly as it went. The subconscious lead I had put on the throw was perfect. The bear was five to ten feet away and closing. I watched the black box spinning on its trajectory. My feet began to move sideways as the bear kept coming. The case slammed into its snout. It was a direct blow, square on the nose. There was a sharp thud of weighted plastic hitting fur and flesh and bone. The bear gave a deep, surprised groan. Its head turned away, driven by the angle and force of the impact as the case bounced haphazardly over one charging shoulder. The great head was now turned almost ninety degrees to the side, with my flailing body in the utmost periphery of its vision.

This was my chance.

My body screaming in high gear, I leaped right, staying on my feet. As I darted sideways, the massive head swung back around, once again able to see what was in front of it. The momentum of its charge carried it past me, and I barely cleared the way. My chest shook with the reverberating growls. It twisted its head, looking at me through dark eyes as I shrank to the side. Its left forepaw whipped out in a wide, fast swing as it passed me. I saw the dark callus of its paw, and long, curving claws. The paw blurred past, missing by a foot or more as I kept backpedaling. The entire time, it grunted and growled. My one bit of offense, the Pelican case, was gone, and I was now 100 percent on defense. Weaponless, my hands were wide, ready to palm away a glancing blow or latch on to something. The swinging paw reached the end of its arc and planted firmly, perfectly, back onto the granite of the dome. As the bear passed, I turned, squaring my shoulders to it.

With surprising speed, its charge shifted. It dug its paws into the ground several steps from where I stood, and turned back at me. There was no pause, no respite—the bear's change of direction was instantaneous. As if taking a sharp curve on a banked track, it was coming at me again, instantly back up to full speed. The ground thundered. "No!" I said, as much to a higher power as to myself or the bear. This was not how this was supposed to happen. I didn't want to die. I wasn't ready to die.

With extraordinary speed, it closed the space between us. The wide head and broad shoulders, though daunting, were nothing compared to those piercing eyes and growling, snarling jaws. I stood low, feet wide. In a heartbeat, it had reached me again. I readied myself to leap aside at the last moment. My brain, all focus, processed the frightening speed of the attack. The flash of brown was only a step or two away from me now, and I pushed off my left foot, again dodging right. The animal's momentum kept it on its track as I leaped clear. The jaws snapped, but I was out of reach, and they closed on air. At the same time, it swung its forepaw in another sweeping arc. I twisted away. My left shoulder, now the closest part of my body to the bear, caught a massive dark mitt full of claws. A glancing blow, but I felt the pressure as the claws ran from the point of my shoulder down the meat of my triceps before its continuing arc and my movement pulled them away.

I paid no attention to the swipe. The arm was still functioning, whatever the injury. It was the least of my worries. Still in the motion of dodging the blow, I twisted around to face the grizzly. No sooner had I turned to see it than it was coming at me again. With heavy grunts and thunderous growls, it charged again. Its focus was even more intense this time, if that was possible. In a few pounding steps, the grizzly was back within arm's reach, and I leaped to the side once more, again twisting to avoid the reach of teeth and claws. I turned my torso away, and claws raked across my back. The sensation was immediate yet also distant, my body automatically filtering it out from my direct attention but unable to numb it completely.

Eyes wide, gasping for air and yelling for the attack to stop, I turned once more to face my assailant. Each successive charge had brought it closer to me. Each time I turned, I looked at a bear fiercer than in the last face-off. I was looking at a grizzly so close to me now, it would take only one step to reach me.

It took that step. Once more, angry jaws snapped, and the huge paw searched the air. I twisted and lunged away again, earning more stinging stripes down and across my back. As the claws finished their run down my spine, I whirled around to face the bear again. And just as quickly again, it had turned around. I was running out of space, out of time. There was nothing between us but air, and precious little of that. Everything was moving so fast.

It charged again. Short, hard steps, and it was to me. No room, no time to jump. A flash of ivory teeth behind ragged lips as angry jaws opened. I could have touched them. The bear's head turned toward me. It was so close. Jaws wide, aimed for my left thigh. Twisting, jerking, I pulled my leg to the side as they snapped shut. An audible clack as the teeth came together in the cold air not two inches from my leg. I kept twisting, pulling away, but there it was. The bear was about to win.

Still just inches away, the bear's momentum pulled it forward, and I watched as the head, jaws still clenched, passed my body and began to turn once more. Its shoulder was almost to me now. On instinct, my body was getting set for the next pass. But there was no time. I couldn't dodge fast enough for this one. I'd reached my capacity for outsmarting and outmaneuvering the bear. Still a foot away from me, the passing brown wall of fur was like the hull of a ship.

Then I saw the bear's arm. In a flash, it was cocking back to swing. To my left, the head was turning. To my right, this arm, this leg reaching. The next instant, it was a blur and suddenly larger. Before I could process, register, and react to anything, I was staring at the open palm of a grizzly bear. I saw the spongy black pads and the deep furrowed lines in those pads. It looked dark, deep, and huge, somehow humanlike. Though it was in rapid motion, the image of that huge palm seared into my memory like a photograph. The paw, poised just inches from my face, flashes bright white before fading back into an enduring image of impending doom. Beyond that dark palm is a halo of fur, and beyond that the long, terrible curve of claws, outstretched and separate. One, two, three, four, five claws. Seen from below, a human hand; from all other angles, unmistakably bear.

The crescendo of sound and energy and intensity had reached its peak. The grunting, the scraping, the yelling, the growling, the dancing and dodging—everything rose into this sudden, unbearable din. And then sudden silence.

The huge paw smacked the side of my face with more force than I had ever imagined possible. The slapping thud cracked across my bones. The impact was like being hit with a board wrapped in leather and driven by a hydraulic arm. It was fast, relentless. The paw lost no more momentum hitting my face than it would have lost batting a bug from the air. Everything echoed in sharp staccato. I grunted with the impact, my head careening to the side from the might of the blow. Now I had felt the sheer power of the creature. It was huge, fast, and impossibly strong.

With that blow, I realized that there was nothing physically I could do to beat the bear. This was not a fair fight. I was about to die. My brain was screaming, but my body had lost the capacity to move. I was not ready. While my head was still twisting and rolling to the side, my feet came off the ground and I tumbled to my left. I was still exhaling. Everything from my right ear almost to my nose erupted in fiery heat as blood rushed there.

My arms were still out at my sides, wanting to do something to contribute but unable to. They folded against my body as I tumbled sideways.

This is it. I'm not ready. What else can I do?

I was still falling, halfway down to the ground, and nearly parallel to it, when the bear's paw got me again, this time closer to my waist. It stopped my sideways motion and threw me straight down to the granite, hard onto my tailbone. Before I could react to the sudden change of direction, the violent slam onto bedrock, the bear's head was at my thigh. I was faceup, still tense and sitting up. The huge head obscured most of my leg. Its mouth was open. There was no dodging it this time. I was about to die. Such sorrow and loss. The parted jaws closed over my right thigh. Sudden, screaming pain that adrenaline did nothing to soften. Agony invaded every sense, every nerve and cell of my body. Involuntarily I yelled. The awful pressure of teeth, the tearing of my flesh, the rending of muscle.

I'm going to die, I thought. It was so wrong. The pain filled everything, and then suddenly, it hit my kill switch. Midscream, midflinch, midbite, it all stopped. In that instant of the bite, everything turned to black.

Silence, then nothing.

It is as if I have entered what the Tibetans call the Bardo—literally, between-two-existences—a dreamlike hallucination that precedes reincarnation, not necessarily in human form.

—Peter Matthiessen, *The Snow Leopard*

CHAPTER THIRTEEN

Awakening: Day 29, 19:31

As if I were waking from the blackest sleep, layers of fuzzy consciousness shone through the black of my perception. I felt the pull of gravity. The universe spun around me on multiple axes. I felt as if I were tumbling under a wave in the deepest ocean. There was no up or down, and I spun freely between them, trying desperately to discern one from the other. If I swam in the wrong direction, I would only descend deeper into the blue shadows, sealing my fate. If I found the true way out, in a few short strokes I would be breathing clean air once more. The spin and roll continued as my still-blurry vision brightened. The spinning was disorienting, nauseating, confusing. Where was I? A glimpse of my skin slowed the rotations for a moment before it let go. *My hand.* The other hand appeared, and now both were in sight. Focus pulled close enough for me to see them clearly before the spinning resumed. I saw sky, then earth, and back again, like a whirling globe. As my vision came back around again to my hands, this time it latched on. Left and right eyes dragged together until one image of two hands formed. They were palm down.

The fuzzy glow cleared, and the stubby grass and pink granite beneath my palms racked into focus. *What* ... I was facedown in the fetal position. *What happened?* I was on the brink of losing grasp again. The edges of my vision darkened and blurred, starting to roll. *I'm alive,* I thought—more of a realization than a proclamation. Even my thoughts were out of breath. The ground tilted first one way, then another, like a marble maze. I could still see only my hands and the pebbles and moss immediately around them. *Focus.* I still had no idea where I was, what direction I was facing, what had happened. But with the realization that I was alive came a sudden surge of exhilaration. I hadn't died. And I hadn't woken up in the midst of a terrible mauling.

Mauling ...

My hearing hadn't yet returned, but everything seemed quiet and still. In the brief moment since I'd begun to see again, the only movement had been from inside my head, which was still whirring. My focus kept pulling back to blur every few seconds, but it was steadily growing clearer, as was the contrast of my hands against the ground. Suddenly, the rotation of my brain around my body clicked into place, like a heavy lazy susan hitting a detent in its spin. I knew which way was up and how to tilt my head there. I lifted my head, watching as the barren ground spread out before me in sparkling pink, gray, and waving green.

Pop! There went the horizon. It was tilted awkwardly, thrown off almost forty-five degrees. My roll had yet to be aligned. With the horizon in sight, though, my brain quickly reset it to level. With a strange suddenness, my orientation was set, my hearing clicked back on, and my vision cleared.

I was still on the ground, in the fetal position and not sure where I was or what had happened, but the smile I had inside could have outshone the noonday sun. I could not believe I was alive. I'd been sure I was going to die. With my eyes locked on the horizon, I began scanning right, in the direction I thought the bear had come from. I hadn't moved my head more than a few degrees when a brown form appeared at the edge of my vision.

The bear was still there. I froze.

Adrenaline flared once more. I felt my senses come to a razor edge, my vision sharpening and widening. Cautiously now, ever so slowly I turned my head, leading the turn with my eyes. The bear was some fifteen feet away. I turned until I could see the entire animal, read its motion. It was moving quickly, at a hurried lope. It was intent. Was it coming to finish the job it had started? Was my elation premature? No. The bear was moving away from me.

Could it really be running away? Its heavy strides were lighter than before, as if it were trying to disappear unnoticed. The gait was smoother this time, too, less up-and-down. With each decisive step, though, the bear kept its eyes on me. Its head was turned sharply over one shoulder, watching me as it cantered away. I was down on the ground, immobilized, eliminated as a threat, and this bear was set on my staying that way. I was not about to argue the point, so I held perfectly still, watching it through the edge of my vision to avoid eye contact.

Every couple of steps, the bear swung its head to the other side to continue watching from there. It was no longer grunting—not making much noise at all, in fact, except for the occasional scrape of claws against rock and tundra, but even then, hardly at all. It was almost as if the bear were tiptoeing away, suddenly weightless. The look in its eyes, though, when it turned its head, was intense. If I were to give it an emotion, it would be a mix of fear and anger.

Regardless of the bear's face, its body broadcast its intent. It wanted to leave, quickly. And I wanted it to. My pulse was once again racing, and I focused on staying motionless. *Don't come back, don't come back, don't come back, don't come back*, I thought, hoping my thoughts wouldn't jinx anything. Swiftly and silently, the bear reached the top of the granite dome. With each great swing of its head from one side to the other, it locked on to my motionless form. It crested the dome and was soon dipping below it on the other side. I watched the last turn of the head and the last glimpse of brown fur.

In the silence, the cicada-like drone vibrating through me was deafening. The bear was gone. It had run away. I waited several long, painful seconds, giving it more time to increase the distance between us, so that when I stood up, it wouldn't be able to see me over the rise.

Quickly and carefully, I pushed off the ground with my hands and rolled up to a kneeling position. *Don't come back.* My adrenaline was flowing heartily still, and most of my wounds were dull, distant aches. *My leg. What happened with the bite?* I remembered it clearly, but at least for now, the sharpness had dulled. I hadn't seen any blood, but I had to be bleeding.

Still watching the dome and wanting like hell to get out of there, I put my right hand on the thigh. I felt wetness on my pant leg. *Shit,* I thought, *if there's already this much blood, I'm done.* A quick evacuation scenario played in short images through my head. With a massive bleed, I'd get back to camp and crash right there. It would be hours before transport could arrive, if then.

I pulled my hand from my thigh. It was already sticky. How long had I been out? The stickiness wasn't like drying blood, though, and I pulled my gaze away from the sky and stone for a moment and looked at my spread palm. I expected the red-brown of thick, dirty blood. Instead, stretching between my fingers and in a long, shimmering strand that draped down to my thigh was the slimy spittle from the bear's mouth. *Not blood!* I glanced down at my khaki pants. No blood there yet, either. *Get up!* I stood up—not completely, though, and with my eyes fixed on the granite dome between the bear and me. My right thigh, still numbed by the adrenaline, blossomed with heat. Ignoring it, I stayed in a low crouch on the off chance the bear was watching from a lower vantage point to see if I would get up.

I just got mauled by a grizzly bear, I thought. *I need to get back to camp, and I need to do it before the adrenaline wears off.*

The distant pain in my leg was already beginning to break through the adrenaline. Not far from me, lying on a mat of moss and flowers, was the bright red cover of *The Liars' Club.* I grabbed it and stuffed it in my pocket. Just beyond lay the Pelican case. I limped over to it as quickly as I could.

With the case in hand, still watching the horizon, I started backing toward the ridge and camp. After a few crouched steps, I stood up, keeping low for balance. The granite and grass seemed to move around me as I went. *Don't come back*, I kept thinking. Just as the bear had retreated silently, so did I. I wanted it to have no clue that I was still alive, had regained consciousness, and was hightailing it back to camp. I would not yell, would not cry for help—not yet. Nearly halfway back to the ridge now, I turned toward camp and went as fast as I could. My leg wasn't working as it should, though, and lagged behind the rest of me. I limped along as best I could. Terrified the bear would return, I kept looking over my shoulder, half expecting to see the brown form coming back to finish the job. I had survived one bout with it. A second round would end much worse. *Oh, God, don't come back*, I thought, the plea manifesting as a barely audible whimper through my heavy breathing. I focused on getting away. *Don't trip, resolute steps, keep going*. Camp and support were getting closer with each step. *If I can just make it back to the guys ... If the bear does come back, they'll be able to ward it off. They can help me. They have mace, bear poppers, the sat phone ... That is my job—to get back to camp. I just need to get there, and no one else can help me do it.*

My sluggish leg turned my run into a sort of uneven gallop. Thankfully, I wasn't yet feeling the full pain of my wound. Though still masked by the adrenaline, it was popping through like a cold draft of wind through a crack. I ignored it. I had to get to the ridge, then down to camp. The bear could come back over the top of the dome at any moment, and if that happened, I couldn't be up here. Snapping my head around from one shoulder to the other just as the bear had done, I scanned for it. The domes behind me were clear. The sky seemed darker than when I'd left camp, and though my senses were sharpened, everything seemed flatter. As I ran, the wind picked up. It felt as though it had died during the encounter and was just now picking back up.

There, in the distance in front of me, was the lip of the ridge. Beyond it—if I'd been running in the right direction—would be camp. I had come back at an angle, trying to approximate my slanting path up while taking the least menacing line. In a moment I would find out whether I'd been right. With the last few meters to the edge approaching, I checked behind me again. It was clear. In my mind, though, the bear was right there at my shoulder, breathing heavily. Eyes forward now, I took the last few steps to the ridge. Below me, the lake opened up, and then the shore of the great island appeared. The next steps peeled back the ridge to reveal our canoes! The bright-red boats were like homing beacons. I was on the right path! The rolling slopes up from shore appeared next, then the red and ivory of our Prophet tents. Finally, when I reached the edge of the ridge, almost straight below me was the white cube of the bug tent, where the rest of the guys were still sitting. I could see them, their hazy forms behind the mesh. I was still alone, though. They didn't yet know I was there, and I was still far from them. I could see them now, and if they were to look up, they could see me.

Now, if I yelled, they would hear me.

When I return will it be the same? Will I be the same? Will anything ever be quite the same again? If I return.

—**Edward Abbey,** *Desert Solitaire*

CHAPTER FOURTEEN

Airway, Breathing, Circulation: Day 29, 19:41

"Bear!" I croaked.

The silence, the empty rush of wind across scrub grass and stone, was broken with that one echoing word. I was nearly shaking with the possibility that the bear would appear behind me once more. I checked nervously over my shoulder as often as I could, but now I was about to step off into the ridge. The descent would take every bit of my focus.

"What?" they called up, incredulous.

"A fucking bear!" I bellowed, starting down the ridge. I didn't give a damn if they understood me that time—I was getting the hell down from that ridge. The rustling from the tent suddenly stopped as everyone turned and looked up through the mesh of the bug tent. My adrenaline was flowing freely, and I looked at the ground, wide-eyed and unblinking as I tried to navigate the tortuous terrain. Somehow, I was still carrying the Pelican case, and with my hands beginning to support the injured leg, I had limited ability to balance or grab, so I placed my feet as carefully as I could. Each step needed to be firm. I couldn't fall down the ridge now. Little gravel steps gradually grew in height and became uneven as I descended. The grass got taller, to the point that I could trip on it if I wasn't careful. As I stepped carefully down the precipice, I angled left and right to find the easiest route and get to the group as quickly as possible. In my mind, at any moment, the bear would appear on the ridge above.

At this point, Dan's face was visible in the screen of the tent. "Alex, you can't joke about this! Be serious with me," he said in a commanding voice. "Are you serious?"

"Yes, I'm fucking serious!" I sputtered back, almost screaming. *Fucking believe me!* I thought. I was going down the ridge faster now, the bear a dark shadow in my mind, looming behind me. At any second, its head could pop over the edge of the cliff. I didn't need to watch it now, though. The guys were looking up at me, and they sure as hell would let me know if it was behind me. I could focus on my steps and, good God, the pain in my leg. I must have run through my whole supply of adrenaline, because my leg suddenly started to scream at me. It seemed perfectly timed with my steps as the incline steepened into a giant's stairway. At first, I took huge strides down them, but then they became too tall. Still, I watched my feet and the ground carefully, trying not to blink. If I did, I could misstep at a critical moment and go tumbling down.

As I ran, I noticed a flash of color with each step. Red. Paying closer attention, I could see that it was my left foot. Blood. *What the hell?* I watched it with each step, the urge to get down outweighing my desire to survey my wounds more closely. Where the hell was blood on my left foot coming from? I would have expected it from my screaming right leg, but I wasn't aware of an injury on my left side. The red was getting brighter, bleeding actively as I ran. I watched with the next step of my right foot to see if it, too, would flash red. It didn't. But the swatch of red on the left was still brighter and growing more so with every step.

"Are you hurt?" Dan yelled up to me.

He needs to know where I'm hurt, so he'll know what to do.

Breathless, I yelled down to him, "I got attacked by a bear, and I got hit in the face and I don't know if I'm cut, I got bit in the thigh, and there's blood on my foot for some reason andIdon'tknowwhy." The last five words blended together in one hard exhalation as I took another step down. The side of my face was still burning hot from the bear's paw, and I pictured long strips of torn flesh exposing my cheekbone. The growing pain in my right leg was sapping my ability to concentrate on the steps and on cataloging my injuries. I was flexing through a huge torn muscle, and it grew worse and worse, like a deep, slow burn: numb at first, now excruciating. The pain rose as I descended, step by careful step. It quickly spilled over, and my steps began to falter.

The next giant step was a huge one. It was steeper than I could manage, but I didn't have the fine motor skills anymore to climb down it. I stepped off my left leg, jumping out and down onto the next step, landing on the injured right leg. I caught myself well enough but paid the price with a blinding stab of pain from my thigh. I put my hand on the leg in an effort to stabilize or support it. This did nothing for the pain, and after another half step, I grabbed the leg firmly with the other hand, too, still holding the Pelican case. I gripped tightly just above the knee, wrapping my hands around the entire circumference while the case dangled from my flexed fingers. This stabilized the leg and allowed me to support and move it aided by my arm muscles. The agony grew until flexing it was too much to bear, and I resorted to moving my leg solely with my arm muscles.

Dan was now climbing the bottom of the ridge toward me. "Where is the bear now, Alex?" he called up to me.

"I don't know!" I yelled back, indignant.

"Alex," he boomed back, calm, straightforward, and commanding, "I need to know where the bear is now."

"I said I don't know!" Even without flexing my leg muscles, the pain booming from my thigh was almost all-encompassing.

"Alex, it's important I know where the bear is. What was it doing when you last saw it?"

"It was running."

"What direction was it going?"

"Away," I said, "back away." *Would you fucking listen to me?*

"Away from where?" he shouted. "What direction from here?"

"It was running in the opposite direction, away from here." I felt as if I were on *Jeopardy* and had forgotten to frame the answer in the form of a question. He needed to know whether there was still a threat, but I couldn't understand what the hell his problem was. As I ran down the slope, my view of my feet, the ridge, the tents, and the lake gradually shrank to a narrow band that went from my footsteps down to Dan, my goal. My steps were slowing, though. Nearly three-quarters of the way down, I was burning the last bit of my energy. *I'm so close*, I thought through a grimace as heat shot through my leg like a splash of magma. I was almost down to them. So much for getting back to camp before the adrenaline wore off. Close enough, though.

Dan was nearly to me now—a good thing, for I didn't have many steps left in me. His long, even strides quickly closed the distance between us. He put my arm over his shoulder, supporting my injured side. I hobbled and hopped as he helped hold me, and we traversed the last of the ridge. The rest of the guys were coming up now, all from different directions, each one bringing some sort of supplies. One of the guys laid down my sleeping pad, another brought the med kit, somebody had the bear mace and bear poppers, and somebody else the satellite phone.

Dan and I hobbled over to the foot of the bright-orange sleeping pad. He laid me down carefully, awkwardly. I was in agony. All six of us were now together some fifty feet up from the tents. I sat up, supporting my torso with my arms behind me. Dan started to examine me, checking my known injuries as well as checking for others I didn't know about. By this time, my right leg was of no more use than a bone-in ham attached to my hip. Even the *thought* of movement sent shooting, tearing pain exploding through me and out the top of my skull. I was breathless, talking a mile a minute, telling them what had happened. Every few seconds, Dan and the rest of the guys were darting glances up at the top of the ridge. Good. They were looking now, so I didn't have to. And yet I craned my head once more to look at the ridge, unsatisfied they were providing enough of a lookout. It was empty—just a craggy wall with sharp, uneven pinnacles.

Dan had already started his head-to-toe exam. Having checked my head and extremities, he needed to remove my clothes to look at all my injuries and confirm that there were no others, no unseen holes I might be bleeding from. I quickly shed my jacket, shirt, and long underwear. Amazingly, the long johns had only a small tinge of blood around the torn holes on my right upper thigh, where the bear had bitten. Soaking through in mottled circles around the holes were the dark stains. On the next layer, we found the torn tatters of my bloodied blue fortune-cookie boxers, my torn skin visible through the gaping holes in the fabric. Blood was still oozing steadily. Carefully to avoid pulling at the rent flesh, we started to take off my boxers.

"Can you guys give us a little space?" Dan asked.

They spread out and formed a lookout just below us, behind Dan. I held myself up as best I could, and Dan pulled the boxers free. They dragged small rusty smears down my legs.

"Didn't you hear me screaming?" I asked.

"No," he said. "The tent was flapping so loud with the wind, we couldn't hear anything."

With all impediments gone, the blood oozed freely from my wounds, spilling down and filling the dimples of my sleeping pad. At the top of my thigh, just below the fold of my groin, was the torn flesh and bright blood of the bite. Dan pressed it with gauze. He looked at it. My skin was ripped in a ragged tear about an inch square. Blood pooled quickly there and kept dripping over the sides, though the cuts themselves were mostly superficial. At the corner of this ragged square was a dark crater. Blood filled this opening—a deep red pool with several thin tendrils of torn flesh reaching out from its edges like the dendrites of a nerve cell. The blood brimmed over the edges of the crater and spilled out, dripping down the inside of my thigh to the sleeping pad. A deep puncture. This was where the bear's canine had pushed through my clothing and deep into my leg. It had parted skin and torn muscle. Cleaning further, Dan found that it was deeper than we could really see. It seemed to be the depth of the bear's tooth. That put it at about an inch and a half deep. Torn skin, yes, rent flesh, yes, but slow blood flow. That much was good.

The wound could have been bleeding more, probably should have been bleeding more, but it had struck only the slow flow of veins and tissues. Just next to the opening and a quarter inch from the deep tear, pulsing hard with the flow from my still-bounding heart, was my femoral artery. A quarter inch over, and the bite would have severed it. Instead of oozing venous flow, I would have had arterial spurt and spray. I would have bled out on the run back, if I even made it that far. The wound would have spurted and sprayed my life away. I pictured myself running back to the edge of the ridge, compensating for my blood loss until that precipice, then falling and tumbling all the way down. But this wound was not spraying. The artery was somehow unharmed, pulsing away just beyond the edge of this torn puncture.

"Hold this here," Dan said, handing me the gauze. Wincing, I pressed it against the wound.

"Dan, I've got to ask," I said, still out of breath, "and this is a dumb question, which I'm pretty sure I know the answer to, but … I'm not gonna die, am I?"

"No, you're not going to die," he said, still working on the wound.

The puncture that so narrowly missed my femoral artery was also alarmingly close to my privates. They had shrunk back as if in horror, pulling away from the damaged leg. Luckily, they'd been spared.

We weren't done yet discovering wounds on my leg. Back on the other side of the puncture, the ragged square of lacerations still bled freely. Cleaning them, it was apparent that I had underlying tissue damage beyond the obvious cuts. Below the tears was dark bruising, though I rarely bruise. Only a few short minutes had passed, but already the skin had turned a deep purple mixed with flecks of red—a compression wound. The bear's tooth had pushed here with nearly the same force that had torn my leg just a fraction of an inch away. The tooth had pushed into my flesh with terrific pressure, compressing skin and muscle tissue to the point where they couldn't bounce back. They were damaged, perhaps permanently; blood was not flowing back to these sites. We would soon learn that without the needed blood flow, the tissue in this spot would die.

Dan found four more of these wounds. There was another from the same row of the bear's teeth as the puncture wound, this one pressing deep into the outside of my thigh just at the edge of my quadriceps, and another just down the leg from that, also from the upper jaw. The last two were on the underside of my thigh, from the lower jaw. My entire leg had been in its mouth. The skin and muscle on both the top and bottom of my thigh were in rough shape. Five separate injured sites. No wonder I had so much trouble getting back to camp.

Still breathless, Dan continued to clean and examine my wounds. My feet were bleeding freely but were not an imminent threat to my life, so he ignored them for the time being. So did I. He ran gloved fingers through my hair and over my scalp, looking for hidden wounds. There were none. He said nothing of my face, so I assumed it was free of anything too grotesque. But I had to ask, "Is there anything on my face?" I felt the cool drip of blood down my neck and then my back. "What's this?" I asked, pawing at the red line.

Dan traced it back to my ear. "Just a little cut," he said, putting his fingers to my earlobe.

"Really?" I asked, disbelieving.

"Yeah," he said, already moving on. "Your necklace is gone," he said.

I thought of the thin silver chain a friend had given me. "Oh, shoot," I said with a sigh, resigned to its loss. If there was any acceptable way to lose something, this was surely it. I was lucky not to have lost a lot more.

Dan moved beyond my neck and shoulders, continuing the head-to-toe exam. Parallel groups of thin raised red lines ran in sharp streaks across my back, from each shoulder down and across my spine. Claw marks from when I'd been dodging the charging grizzly. The claws had swiped my jacket, the thick pile of the fleece and tight weave of the windproof fabric absorbing much of the force from the blows. The marks on my skin looked as though they could have been from fingernails, but they were sharply raised and consistent. I could feel them, like hot scrapes.

The fibers on my fleece were flattened, almost burned in places, from the force and the friction. Other than another set of stripes running down the upper half of my left arm, both arms were fine.

Continuing his full sweep, Dan found scuffs and scratches covering the rest of my body, but nothing too major until we got back to my feet. My left foot was still bleeding, and it was time we looked at it more closely. The blood was coming mainly from my big toe and second toe. The tips had been sliced cleanly off, leaving a circle of raw flesh at the tip of each. They were bright red and still bleeding. The slice on the big toe was a good centimeter and a half wide; on the second toe, it was just shy of a centimeter. I had no idea how that had happened. I had no recollection of hurting my foot, and no idea what would have cut them so cleanly.

Dan took off my bloodied Chaco sandals, loosening the straps and sliding them off my damaged feet. As I sat facing downhill, focused on my feet, the distant view of the island slope and the lake beyond was like a blurred backdrop in a photograph. Dan worked his way up from my toes and found the next large cut, a sweeping arc that ran around my ankle joint. This was most likely from a claw, and it formed a half moon around the ball of my ankle. The arc was bleeding, though nowhere nearly as badly as the toes or other injuries. The other foot had several cuts but nothing too major, though most of them would likely scar.

Dan was flying through this examination, and maybe a minute had passed since I reached the sleeping pad. "Go get the big pot full of water," Dan said. Darin popped up and ran to the bug tent, then down to the shore.

> That's not a thing any of us are granted. To go back. Wipe away what later doesn't suit us and make it the way we wish it. You just go on.
>
> —**Charles Frazier,** *Cold Mountain*

CHAPTER FIFTEEN

What Now? Day 29, 19:45

I was still worried about the ridge. The guys were watching it, but I felt as if it were watching me. I felt its presence. It seemed like a veil, hiding something that was just waiting for me.

I thought back to the attack and shuddered.

For the first time, the gravity of what had happened hit me. I had been mauled by a *grizzly bear.* I had been sure I was about to die, and nearly did. And now here I was, having my wounds looked after, and likely to *live.*

But we were still in the middle of nowhere.

The edges of my vision darkened, shimmering as my visual field narrowed slightly. The terrifying memory flickered in my head now in sharply vivid moments. Hard, loud sounds and intense, violent flashes of movement slammed into my psyche. *I nearly died*, I thought again, my breath speeding up. *One more quarter of an inch, and then …*

With this realization—one based on evidence rather than on the previous visceral images—my peripheral vision shrank. The edges of darkness closed in around me, and its shimmering borders crept toward the middle of my consciousness. I was starting to hyperventilate, too. *It's not just about me*, I thought. *This is about much more than me. This is about me and ALL the people I know, and some I don't even know.*

My narrowing vision held steady now. It was like looking at the world through a hole in dark paper. This affected many people, and my death would affect many more. My dying would be less about me than about everyone else. I didn't want to die, but if I did, the ones hurting would be the ones left behind, the ones reeling from it, the ones carrying it with them like a dark, violent blot in their memories, for the rest of their days.

With the start of this realization, an image of an old-fashioned Rolodex popped into my mind as if being presented to me. Instead of names and numbers on the cards, there were faces, portraits like those from the high school yearbook, with a head and shoulders in front of some nondescript backdrop. This was my Rolodex. These were people I knew, people I cared about, people who I thought might care about me. Just after appearing, the first card flipped down to reveal the second. It was another portrait, another friend. Then, more quickly now, the second card flopped over to reveal the third. On they came. There was no logical order to the faces, but they were all people I knew—friends, classmates, family. Cascading, accelerating, each card and its portrait appeared for less time before flicking to the next. Ten faces, fifty faces, a hundred. The cards started to blur, though each face stayed for a frame, each one distinct from memory. As the Rolodex accelerated, my vision narrowed and my breathing spiked, and I began to reel. I felt my arms start to give out, and the horizon behind the portraits started to bend. It was too overwhelming. This was too much.

"Dan," I said between breaths, "I think I'm blacking out."

"Okay, lie back," he said, helping me ease down onto the pad. "Just stay with me. Try to keep talking."

I put my hands to my forehead. The ground shifted around me in slow, heavy undulations like the waves in a waterbed.

Darin walked up, hindered by the enormous pot full of drinking water from the lake. My mind flashed to memories of cold trout fillets sloshing in that same pot. Now it would be employed in cleaning the flesh of my leg. He set it down near me, within arm's reach for Dan. The heavy, dented pot clanged on the stone as water sloshed over the edge. "Need anything else?" Darin asked.

"Not right now," Dan said without looking up. Darin glanced briefly at me before turning back away toward the bug tent, where the rest of the guys were congregating. Dan looked through the med kit, peering through Ziploc bags until he found the one with wound-cleaning supplies. He dipped gauze in the pot of water and dabbed at the blood around my thigh. In a few short swipes, the gauze was saturated, and he turned a fresh section to my leg. Before long, this, too, was filled with blood, and Dan dipped it into the pot. The water darkened to a rusty brown, and he squeezed the gauze, spilling color into the pot. The rag, restored to a dull pink, was ready for more.

With the wet rag, Dan wiped dried blood from around the wound. Soon, there was no blood except what oozed slowly from deep in the wound. Occasionally, it would bloom up over the edge, and a wayward drop would run down the inside of my thigh, onto the orange sleeping pad. With the blood gone, it was easier to see the lacerations, the heavily bruised tooth marks, and the deep, dark puncture.

I had various little bruises, scrapes, and rivulets of blood all over my body, many of which we couldn't figure out the exact cause of. The injuries to my face were obvious. When the paw had struck me, it just missed my eye. The brunt of it hit a bit low, and the distance was precisely the right length to allow the meaty pad of the bear's paw to connect with my sturdy cheekbone and for the claws to surround my face without touching it.

One of the guys brought me my folding compass, and I examined my face in the small sighting mirror. My earlobe, where the one claw had caught me, was still dripping blood down my neck and back and would slowly coagulate over the next six hours. Where the palm pad hit my cheekbone were two bruises and corresponding scrapes, caused by the bone structure in the bear's paw, where the metacarpals had transferred the force to my skull. One of these abrasions ran directly underneath my right eye, along the top edge of the cheekbone. Around the outside of my eye socket were short scrapes that ran horizontally for a half inch. It still felt as if my face had been cut, and it was throbbing. My cheekbone and the space under my eyelid were bruised and turning into a black eye. I could feel the tissue starting to swell—just a subtle, building pressure along the bottom and corner of my eye.

My leg throbbed steadily, and stabbed with pain whenever I moved it the wrong way, but overall, the pain was beginning to numb as my feeling in the area dissipated, as if dripping out with the blood itself. With my equilibrium mostly back after nearly passing out, Dan handed me another pack of gauze and had me hold pressure on my leg again. The pain from the pressure made it so that I could push only so hard before I had to back off. While I held the direct pressure, he worked on the other wounds around my body. He cleaned my toes and put bandages on them before wrapping them with white tape. Given the number of cuts and the size of our med kit, there were many smaller wounds that we had to leave undressed.

"Do you want me to take any photos?" Dan asked.

I was surprised by the question, "What?" I said, almost indignant. "No." It was far from my mind, but a logical enough question, especially given my love of photography.

"Okay," Dan said, returning to his work, "I didn't know if you'd want one."

Since sitting and lying on the mat, I'd noticed a slow ache beginning to radiate from my tailbone. It slowly got worse until sitting a certain way sent sharp pains up from my backside. I mentioned it to Dan, and he checked again, finding no surface damage. It felt like a bruise, and a bad one. I added *bruised tailbone* to my running list of injuries—still not bad considering I had been attacked by one of the earth's largest carnivores.

There. It hit me again—the enormity of being attacked by a grizzly bear. "I've got to lie down again, Dan," I said.

As I lay back, my thoughts spinning, Dan finished cleaning my leg, wiping up the fresh blood with a dry piece of gauze before dressing the wound. Over more gauze, he wrapped an elastic bandage. The location of the wound made dressing it difficult. It was essentially at the top of my leg, in my groin, involving the hip girdle as much as the thigh. A wrap just around the leg would ball up and cut in, but the Ace wasn't long enough to wrap properly around my waist and then my leg. He turned the wrap around my thigh in wide, even bands, pulling the wrap at the outside of my leg to keep the bandage from bunching up into a thick, constricting rope. We used as much from the med kit as we dared. Reaching the end of the bandage, he secured it with the tiny clips. Dan closed the med kit, then looked me over once more, reviewing his mental checklist of my wounds as he eyed the bandages.

Satisfied that he'd checked me over thoroughly and that the wounds were bandaged up as well as they could be with what we had in our med kit, we worked out how to move me. Mike had gone to the tent and fetched a fresh pair of boxers for me, as well as fleece pants and several other articles. The long johns I'd had on were still in good shape and largely free of blood, so I could continue wearing them. Their only visible evidence of the attack was the small pencil-width hole a few inches down from the waist and almost to the inseam. The bloodstain had dried, turning even darker, like spilled ink. With a good deal of help, I was clothed again. Dan carefully fed socks over my thoroughly taped feet and toes.

Slowly coming to my senses and feeling the air and space around me in a calmer way, I started to take in more of my surroundings. I realized that the hat I'd been wearing was gone. A flash of the huge paw just before it hit my face flickered in my mind. I pictured the hat coming off with that blow, knocked free as I was thrown sideways. I closed my eyes in an effort to turn off the vision.

"Damn," I said through a sigh. "I think my hat's up there."

"Are you saying you want us to go up there and get it?" Dan asked in disbelief.

That would be nice, I thought, but I would never have asked anyone to go.

"No," I said. "I just realized it was gone."

With help, I put on several layers over the top of my body, finishing with my puffy down jacket and black Patagonia fleece cap. I was covered nearly head to toe, with just my hands and face exposed, the beginnings of a black eye and my obvious disability the only evidence of my ordeal. I couldn't move my right leg. Any motion involving the greater muscles of that limb nearly laid me out again. Any slip or accidental twitch—even an errant thought—could send pain like a twisting knife through my thigh. That muscle was attached to much more than I had ever imagined.

Figuring out the logistics of my movement was the next big task. Somebody grabbed my spare paddle, the pathetic brass-colored aluminum-and-plastic one. It was longer than the rest of the paddles, but more importantly, we didn't care about it. It just might suit as a crutch in a pinch, and this was a pinch. Jean and Auggie brought the paddle and rigged more long underwear from my bag as a pad. We tied it to the handle end. Several of the guys gathered around to help lift me up for the trip to the tent. I couldn't help. My feet were out in front of me, and I had to get off the ground without moving one of them—not an easy trick. They surrounded me and lifted my torso until my legs were under me. Several times, either I twitched or the maneuver shifted my thigh in the wrong way. I gasped from the pain. This needed to happen. I couldn't just continue lying on the slope. No matter what, I had to move.

Grimacing, I got upright, and with Dan under one arm and the crude crutch under the other, we moved slowly toward the bug tent. While we hobbled along, the paddle digging into my armpit, the other guys moved my sleeping pad and all the gear spread around our aid station to the tent. Before we'd made it halfway, my station was set up in the bug tent. I hobbled like an arthritic old man with a bent cane. It was exhausting.

Eventually, we had reached the tent, and the guys held up the skirt of mesh as high as they could. I still had to bend and almost crouch under it to get in. We awkwardly clambered in, and with help, Dan lowered me to my waiting pad and sleeping bag.

Situated, Dan asked how I was, and I told him I was fine.

"All right, guys," he said, more to the rest of the group than to me, "I'm gonna call Camp now."

Phone in hand, he walked away. Just beyond the shelf that housed the bug tent was a steep drop down to another plateau. In a few steps, he had nearly disappeared below the rise. Crossing his arm, he turned his back to the tent and the wind, shielding the phone from the gusts as he waited for the sound of a ringing phone on the other end.

The last tie with safety was being broken. More than three hundred miles of practically unexplored wilderness lay before us, down a river traversed, perhaps, by only a handful.

—**Eric Sevareid,** *Canoeing with the Cree*

CHAPTER SIXTEEN

The Evacuation Algorithm: Day 29, 19:50

With Dan gone, silence filled the tent. We all were unsure what to say. The stove's constant roar filled the background with white noise as it cooked a pizza in the outback oven.

"What do you want on your pizza?" Darin asked, breaking the silence.

"I don't care," I said. "Whatever." I wasn't hungry.

Silence returned to the tent. In my mind flowed an undercurrent of worry that perhaps, at any moment, the bear would return. But whenever the worry left even for an instant, I was filled with elation. I was so happy to be alive! I didn't care what pizza I had, or that I was a little uncomfortable. The pain, soreness, and discomfort were a small price to pay for being alive. In this state, though, I didn't know how to talk to anyone, and they didn't know what to say to me.

The tent was small, but it didn't feel crowded with the five of us in there, even with me lying down on my sleeping pad. Around me were four guys, four walls, and a roof. It felt like the first bit of safety I'd had since the attack. This was partly irrational, though—this thin layer of nylon would do nothing to protect any of us if something wanted to do us harm. On the other hand, there truly is safety in numbers. But always in the back of my mind was the island where we sat, the boundaries of its shores, and the grizzly bear that was still here, somewhere. I tried to push the thought away, to find solace in the bear mace, the bear poppers, the group, and the phone Dan was on right then.

I was exhausted. As I shifted my weight and felt a stab of pain, I suddenly realized I had become completely dependent on everyone around me. Eventually, Auggie spoke, and his question pierced the silence cleanly, driving straight into the heart of it.

"So what happened?"

They had been away from Dan and me most of the time he was doing patient care, and hadn't heard my breathless account. They all were still mostly in the dark.

Dan waited as the sat phone rang. He needed to talk with someone, to tell them what had happened, to bounce ideas off them. The phone rang again. Any moment now, someone would pick it up. On the caller ID, it would just look like an odd number, a foreign caller. He wasn't dialing a special line. This call rang the same as all the rest. But no one was there to hear it. The phone stopped ringing, and a recorded voice came through the tiny earpiece. *"Thank you for calling Camp Menogyn ..."*

Dan was startled to hear the answering machine. He'd been expecting a person, expecting to be able to gush out whatever came to mind, and let a calm, understanding, rational mind on the other end interpret the tumbling fountain of words. Instead, he would soon hear a beep and then have a short window of time before he was cut off. *Oh, shit,* he thought. *How am I going to leave a message? What do I need to tell them? How do I do it in a voicemail?*

"... but if you leave a message with your name, phone number, and a brief message, we'll get back to you as soon as we can. Thanks!" The recording ended abruptly in a sharp beep, and the line went silent.

Back in our flapping bug tent, I looked around at each of the guys. They waited for the story. Shuffling through my memory, I decided to start from the beginning. My voice began to shake, and my body quivered with an unexpected tremor. I felt weak and terrified as I spoke, as if the retelling might make it happen again. They listened in silence as the stove hissed and I told of seeing the bear, mistaking it for a musk ox, and then everything that followed: the bluffs, the charge, throwing the Pelican case, dodging the bear, nearly getting bitten, getting hit in the face, the pain, and everything going black in an instant. I told them about how I'd woken up, how the bear was still there, and how I'd waited for it to leave, how terrified I was that it would come back, and how I had run limping back to camp. The story hung like a heavy fog in the air, and my mind set to wandering. Some of the guys acknowledged the story, but mostly, they were shell-shocked. I wouldn't have known what to say in their shoes, either.

Dan shielded the phone with shoulders hunched against the wind, and spoke into the microphone to record the voicemail. "Alex got bit by a bear; he's okay," he blurted. "Um ..." He looked down at the faded block numbers on his watch. "It's seven fifty ... I'm going to call back in ten minutes. Can somebody please be there?" He could hear the strain in his voice—stress mixed with pleading hope.

After a brief moment, having said everything that needed to be said, Dan took the phone from his ear and thumbed the red End button.

He closed the large antenna and looked out away from our huge island, at the surface of Princess Mary Lake. *Well, we can't really evacuate,* he thought, looking at the waves and the distance and the steep shores, *and there's a bear somewhere on this island.* With a sigh, he put the large phone into the pocket of his blue Patagonia Gore-Tex jacket before turning back into the wind and toward the bug tent.

"Darin," I said, shaking myself out of the swirling thoughts in my head, "would you take a picture of me?"

"Sure," he said as he stood up and grabbed the case, popping the stiff latches open and pulling the heavy Nikon from its tight foam nest. "What do you want?" he asked.

"Just a picture of me lying here," I said. "Might as well." It felt odd to ask for a photo. I'd been so reticent just a little while ago—indignant, even—when Dan asked whether I wanted to be photographed. In hindsight, I wish I'd said yes that first time. I would have had a photo of those important moments, a timeless image instead of gradually fading memories.

Darin flicked the power switch on the D70. "Okay," he said, putting the camera up to his eye. I turned my smile into a grimace and looked into the camera. He clicked the shutter, exposing the first picture of me since the attack.

I laughed at my facial expression, bouncing right back to the giddy elation of being alive. Not half an hour ago, I had been sure it was the end. I had no idea what would happen to me now—what kind of medical treatment I'd need, what sort of challenges I would face in dealing with the injuries, or how long it would take. None of that mattered. I was alive after thinking I couldn't survive. There is no feeling like it. No matter what pain or struggles were ahead, so long as I could make it out alive, I was up to the challenge.

But I really did want to get the hell off the island.

I was terrified that the bear would come back. To have survived one bout only to be struck down in another—the thought was too awful, the only no-win scenario. It flicked into my mind every few moments—a flash of fear burning in my chest. *But if that happens, this time you're not alone*, I told myself. *You were alone, and you're not anymore.* I realized that my breath was speeding up, my eyes fixed on a blank distance. I blinked and deliberately slowed my breathing. *You are not alone anymore.*

"How's that pizza?" I asked, trying to distract myself.

Auggie peeked under the skirt and lid of the outback oven. "Nearly done," he said.

Dan appeared from his calling spot and told us he was going to call Camp back in ten minutes.

"Can I get some pain meds?" I asked him.

"Soon," he said, "but not yet." He checked on the rest of the group and then climbed under the mesh to join us in the bug tent.

My thoughts spun in the silence until Darin pulled my pizza out of the oven and presented it to me on one of our multiuse lid/plate/trays. What little appetite I'd had was gone again, but I knew I needed to eat, so I started in slowly. The pizza was fine—rather meager, but a meal that not too long ago, I didn't think I'd get to have. I appreciated it, but I didn't really want it. My body was still on alert, my startle response on a hair trigger, with my reserves ready at the twitch of a muscle.

I ate quietly in the hushed tent and felt the slow passage of time.

After a bit, Dan rose and left the tent, satellite phone in hand once more. He turned down the hill and disappeared over the rise.

I'd gotten through maybe two pieces of pizza when Mike, who had gone off to the tents for something, came running back, out of breath. "There's a big animal moving toward camp," he said.

Liquid energy shot through my body like a flash of hot light. With the adrenaline, my breathing and heart rate shot through the roof. My eyes widened. *It's back!* I thought. *No!*

My back was to Mike and to the large animal. "What!" I yelled, my voice revealing my horror, "It's *back*?" I pictured the bear roaring through, destroying our tents and marauding the camp before coming, with a painfully slow inevitability, to finish the job it had started. I twisted on my sleeping pad, craning to see what was going on, trying to grasp what I could of what was happening, and at least get my head into the game, even if my body had to sit it out. Twisting hurt, but the surge of adrenaline masked a lot of it. I scooted my hips back with my hands and pushed myself up so I could twist better, and turned so I could see Mike. He was still veiled by the mesh of the tent, and the blurry distance behind him even more so.

"No, no, oh, God, sorry," Mike said quickly. "Not a bear."

"What?" I asked.

"It's a musk ox," Mike said.

"Oh, Jesus!" I gasped, my head swirling. With this new knowledge, though, I felt the first bit of calm, starting at my brain and trickling slowly down as the adrenaline in my blood began to ebb. "You can't fucking do that!" I growled at Mike. "I thought it had come back … I thought it was the bear …" I felt as if the earth could start spinning again.

"No," he said quietly, his hands open and patting the air, as if to calm me down—or maybe push me away should I lunge at him, "It's just a musk ox."

"You scared the shit out of me," I said, still reeling. There were more choice words in my mind, stronger words that more accurately expressed the depth of my horror and anger, but I was exhausted. Relieved that it was not the bear, worried that it truly was a musk ox, and dismayed at my reaction—a reaction I was not in control of—I crumpled back onto my elbows on the sleeping pad. My heart still thumped like a jackhammer as I closed my eyes and tilted my head back, as if this might help the adrenaline evaporate out of my body. It didn't. It lingered like a strong alcoholic buzz when all you want is to feel normal again. It stuck there, that whine and tingle of uncomfortable alertness, and I opened my eyes.

Everyone in the tent had stood up, pots and pans in their hands. Then, almost in unison, they stepped to the nearest gossamer curtain, lifted it from the tundra floor, and left the bug tent.

"Wait!" I cried, reaching out and grabbing Mike. "Do NOT leave me alone here!" My adrenaline spiked again. I was about to be alone in the tent, injured and vulnerable, as a potentially dangerous animal was sauntering into camp, and as another, its appetite freshly whetted, lurked somewhere in the unknown shadows of the island. Mike had turned to leave, to help the others ward off the musk ox, but he turned back to me. He had to see the wildness in my eyes, but there was also a pleading in my voice, and worry etched on my face.

"Okay," he said, and he stayed. My worry abated just slightly, and I tried to let more of it out with a long, slow exhalation through parted lips. Most of the tension stayed, but some of it drifted off with the wind.

Turning awkwardly around on my sleeping pad, I could see our tan sleeping tents in the corner of my eye. Their domes dotted a rise a hundred yards away from the bug tent. At this angle, I could just see the tents and the ground beyond them when I turned. In my blurry view, our small force of Jean, Darin, and Auggie was moving toward the tents. To get a better view of what was happening, I twisted farther, leaving my leaden right leg on the sleeping pad. On a far hillock beyond the guys and beyond the tents, a shag of heavy brown fur with great twisted horns appeared over a rise and then quickly disappeared again.

Shit, I thought, *musk oxen are coming into camp and I can't do a damn thing.*

As I turned back, the first loud clangs rang out from our small force as they started banging pots and spoons to repulse the prehistoric-looking beast. I tried to force down the rest of my pizza, though it was cold by now and I was even less hungry than I'd been before.

The three of them stood at the rise near the tents—bright, colorful silhouettes in their rain jackets. Their arms swung out arrhythmically and came violently together. A moment later, the crash of metal crossed the distance, sounding like broken cymbals and wildly out of sync with their movements. They were also yelling at the musk ox. "Hey!" they yelled, "Go away!" But their noise had no effect on the oblivious lumbering creature. They tried other words and more complex sentences, as if trying to reason with it, again to no avail. Still the musk ox plodded on, chomping at the meager grasses.

I was worried. We were still on the island. I was still injured. The bear still lurked somewhere on the island, and musk oxen were coming into camp. My leg had started to grow numb around the bite, and most of the time it didn't bother me, though a dull, persistent ache had been slowly building. Of course, anytime I moved in the wrong way, it was back again—that horrible tearing that shot straight from my leg to my brain and back again like a rebounding lightning bolt. I thought about the bear, about the musk ox, about my leg.

I was grateful to be alive, but far from feeling safe.

Dan had found his spot on the lower plateau and thumbed the satellite phone. It had been ten minutes since his last call. He opened the antenna and punched in the number. The phone rang again, full of promise. After the first ring, the line clicked open. Someone on the other end, in the boathouse office in Minnesota, picked up the receiver.

The phone had rung once, but Aaron, the program director, had been waiting for it. He lifted the wireless receiver from its base on the desk, clicking the line open and connecting the call. "Camp Menogyn," he said. "This is Aaron."

Dan felt a wave of relief. "Aaron, this is Dan," he said.

"Oh, hey, Dan," Aaron said, sounding more casual than he surely felt. "How's it going?"

"Well," Dan started, "Alex got bit by a bear, but he's okay."

"Oh? Is everybody okay?"

If Dan relayed nothing else over the phone, Camp needed to know where the group was. As quickly and efficiently as he could, Dan told Aaron about the island and where we were camped.

"Oh," he said, sounding surprised. He had been to this place and had even advised Dan on the route, this lake, this exact island. He had said to visit there, that the view was amazing. "Sweet site, dude," Aaron said. Dan instantly felt less alone. He was no longer in an unknown place. He looked out at the lake now. It had changed in the past hour from a place of sublime beauty to a place of variables, scenarios, and escape routes.

Aaron confirmed again that everyone else on the trip was all right, and Dan told him that the others were okay. He confirmed that the bear was no longer around and that it wasn't stalking us, as far as we knew. Dan also told him that we were camped relatively close to the water and could get down there in a hurry.

"Can you call back again in about a half hour?" Aaron asked.

"Sure," Dan said, thinking through the logistics of calling back. He looked at his watch, reported the time to Aaron, and added thirty minutes for the call-back time. "I'll call you then," he said.

"Sounds good," Aaron said. "We'll be here."

Dan pulled the phone away from his ear and signed off. He listened to the racket from the musk-ox warding team as they rang across the island in an arrhythmic staccato. It seemed to rush up, sweeping across the island. *Bam! Bang! Clang!* He pushed the banging and yelling from his mind. *Twenty-nine minutes.*

Dan had paced back and forth during the half-hour wait and was back on the phone exactly at the scheduled time. He was keeping his conversation short but was being thorough. On the other end of the line was the warm-hearted Camp director, Paul, and a crisis team. They would know what to do. They would take the information he gave them, and communicate with emergency services, medical offices, corporate offices, law enforcement, maybe even the Air Force. I could see Dan from the tent, slightly blurred by the mesh, pacing with the phone to one ear and a hand over the other to block out the noise. Every few moments, he glanced toward the tent and the guys banging pots and blowing whistles. The lumbering beasts ambled closer while he talked, and he became more and more focused on the conversation.

The entire time, the musk oxen continued toward our site. While not inherently hostile or scary, upon reaching our site and seeing us, they could become deadly. They had just dipped below the rise beyond our tents and were out of view for a few minutes. Suddenly, one of them crested the ridge just past the tents, at a trot. Its huge body was silhouetted against the sky. Its matted fur flowed with the motion, and when it came to an abrupt stop, the mats swung around it like a shawl. It was suddenly close, suddenly in our camp. Startled, our pot bangers stared and paused, then increased their intensity as they backed toward the bug tent.

The musk ox went back to grazing.

Dan worked to finish the conversation with camp, and Auggie, Darin, and Jean backed closer and closer to him, the cacophony of banging pots getting louder with each passing minute. Dan was engrossed in the call, hunched over, shielding the phone from the wind and the racket, his hand still covering his other ear. He hadn't even noticed the pot bangers' approach until he found them standing right beside him. He turned, shielding the phone from the clamor. "Can you guys go bang those pots somewhere else?" he said.

"No, Dan!" they yelled, still clanging the pots. "The musk oxen are right there!"

Dan turned and saw the beasts on top of the hill—three of them now, twenty feet beyond the Prophets. From those tents, it was only another hundred yards to us and the bug tent. They were close.

Dan raised the phone to his ear. "Okay, Paul," he said, eyes on the approaching beast, "I have to go. There's a musk ox in camp."

I watched as, without waiting for a response, he lowered the Globalstar sat phone and ended the call. This was all out of the ordinary, all wrong. In eighty-plus years of operation, camp had never had a bear attack. Only a few years earlier, the long trips had been moved from operating in polar-bear territory after some near misses, only to have the first actual attack happen here.

"All right, guys," Dan said, taking long strides toward the bug tent, "we're evacuating camp. Get the gear and let's go!"

I was still in the bug tent, twisting from left to right to see where everyone was and what was happening. The pot banging had shifted, the sounds moving as the guys retreated toward me.

The commotion escalated around me as guys darted about, banging away on their pots and bowls. The musk oxen were in the camp now, between us and our tents. From this close distance, they looked like huge, horrid puppets with alien movements and fearsome curving horns. I didn't want to be alone. The simple presence of another human was desperately important to me. *It's okay*, I thought. *Mike's here.*

There was scrambling outside the tent, and Mike leaped up. As he slipped under the tent skirt, I yelled, "Mike, come back!" I practically screamed, "Don't leave me here!"

"I won't," he said, looking back at me before he turned and disappeared down a rise.

A switch had flipped in our camp when the first musk ox crested the ridge just past our tents. That was too close. They were still coming toward us, still somehow oblivious to our presence. By contrast, we couldn't have been more aware of them lumbering downhill toward us. I heard the scraping of feet on gravel as the guys ran about. The moving chorus of banging pots rose and fell like distant sirens as the bangers scrambled to and fro, gathering supplies and conferring with each other.

Dan burst into the bug tent, the sat phone bulging in his pocket. He told me we were evacuating to the boats. A few paces behind him was Mike. Dan bent down, and I held my arms out like a pleading baby, bear-hugging him as he wrapped his arms around me and tried to lift my dead weight up off the cold ground. He grunted, and my body lifted a couple of inches before I sank back to the ground. I wanted to help, but there was nothing I could do. My legs hung down like logs on a hinge, doing nothing to help and only impeding. Any motion I made with the lower half of my body sent shooting pains through my thigh.

I had turned enough so I could now see the tents more clearly. At least three musk oxen were now among them. The rest of the guys moved around like skaters on a rink, flowing in great wide arcs around camp while they grabbed essentials such as the med kit before turning down the ridge to the distant boats below.

We really were evacuating camp. This wasn't done lightly. It was windy on the water, and we were leaving our tents, our shelters, our homes. My mind didn't have much time to ponder, but it had been running through the scenario continually as new information came in. Dan was right—anywhere was better than here.

Dan was about to try lifting my body again when Mike caught up with him. They squatted deep and grabbed me as I hooked onto their shoulders. The three of us grunted as they lifted my full weight from the ground. Progress was slow for the first few inches of elevation. My leg throbbed anew, torn tissues shifting with the changing angle. I grimaced as my bearers got past the awkward angles and could put their full strength into the maneuver. Suddenly, I shot up to full standing height, and my legs hung freely down to the tundra. I had an arm around each of them, and I could now hop or at least support some of my weight with my uninjured leg.

From standing height, I could fully see the musk oxen. We were overrun. They were in our camp, which suddenly felt much smaller.

"All right, guys," Dan yelled across camp to the scattered pot bangers, "let's go! Down to the boats!"

Dan and Mike hefted my weight and started down the steep scree and moss toward shore. I helped for the first few steps, each of them painful as my injured leg swung freely. My body tried involuntarily to stabilize the leg as it moved, tightening muscles all around to hold it still. Sudden stabs of pain shot through my body, flashing my vision with white light. I tried to focus on moving my good leg, carefully holding the bad one back with the uninjured hamstring muscles. It shot hot metal through my body.

Mike and Dan's pace quickened, and I couldn't keep up. They suggested I let my legs hang and not try to run with them. I pulled my good leg up, bending it at the knee and holding it back, doing the same with the injured leg. My shoulders burned with the sudden change in weight, but I held. Mike and Dan were suddenly free from the impediment of my help and they started moving more quickly.

Together they turned to face toward camp, taking a quick look at the invading musk oxen, making sure Darin, Jean, and Auggie were long gone on their way to the boats. We were the last three. Several shaggy brown shapes were in camp now, and more had appeared at the crest near the tents. The banging pots had fallen silent as the guys ran to the boats below. There was a sudden calm across the tundra, broken only by gusts of wind, our own heavy breathing, and the scraping of scree beneath boots.

As if noticing the silence, the lead musk ox paused its grazing. It lifted its heavy, horned head from the scrub grass and swung its gaze toward the three of us. Upon seeing the small group of humans, it went still. It fixed on us with a blank stare that conveyed a feeling somewhere between aggression and incomprehension. In this case, we lucked out in that it seemed to lean more toward the latter. The beast's head pulled back slightly as Dan and Mike turned toward the canoes. The musk ox watched. To it, the three of us must have melded into a single tall, multicolored three-headed beast. This yellow, black, and blue man-beast walked—ran, even—with several legs. It must have been terrifying. The creature had finally, suddenly registered our existence.

Every step down the hill was painful, and I struggled to hold myself up and keep my leg neutral. With Mike and Dan watching our steps, I turned my head and watched the musk ox as we continued downhill.

It sprang backward in a surprisingly sharp turn led by its heavy, horned head, and moved with an urgency I had not yet seen in the species. Its dreadlocked hair, which normally hung like a sort of grungy skirt, now flapped up and flowed in quick waves, revealing the spindly dark legs and hooves. It was running away. The other musk oxen followed suit. Like a school of huge, ungainly fish, they turned and ran, disappearing over the ridge in a flurry of waving, grimy hair and clattering hooves, the ground rumbling beneath them.

"Oh, my God," I said between grunts of discomfort. "They're gone!"

Mike and Dan were still running down the ridge, and I turned my head back toward the water. They had been peeking, too, and had seen the musk oxen retreat.

"I know," Dan said, the strain of exertion audible in his voice. "We'll head down anyway."

Halfway to the boats, we regrouped with the other guys. They were panting and a little dazed from the speed of our evacuation, the sudden disappearance of the threat, and the hard run down the hill. Mike and Dan lowered me onto my good leg so I could support myself and they could rest. In Darin's, Jean's, and Auggie's hands was a meager collection of essential supplies to meet our most basic needs. It included our bear-repelling supplies and med kit, along with a smattering of water bottles. They also still held their noisemakers—our pots and pans and metal cooking utensils. Jean held the big pot, the one we'd filled with stews and beautiful fillets and, most recently, water tinged pink with my blood.

"Whoa!" Mike said. "Take a look at that!"

The bottom of the pot had been flat, but over the course of the musk-ox invasion, it had been beaten to a deep concavity like the surface of a steel drum.

Jean held it up, along with the metal spoon he had beaten it with. "What?" he said, a wry grin crossing his face.

We waited there a bit longer to regroup before moving back up the hill. The climb was awful—hard and painful for me, harder for Mike and Dan with me draped over their shoulders. Cautiously, we climbed back up, lookouts scanning left and right of the tents for musk oxen and eyeing the ridge for the brown hulk of a grizzly bear. We weren't in a good position, not having picked our campsite for defensibility. Beneath a steep ridge whose top we couldn't see, we were surrounded by rolling hillocks that concealed any movements outside our camp. Below us was the steep escarpment leading to the water, and our three boats, our lifelines, set onshore with bright blue food barrels resting upright in their compartments. We couldn't leave and could no longer enjoy the blissful ignorance of our surroundings.

We would have to be vigilant until we left, and we were down one able body.

I couldn't help at all. I was a hindrance now, and the reason for all the extra work and fuss. I wished like hell I could contribute. Instead, I was strung between Mike and Dan as they hefted my weight up over shifting rocks and slippery moss.

Eventually, we reached camp again, and my bearers brought me to the tents. They lowered me until I was sitting in the tent with my feet sticking out the door. There, they removed my boots and helped me slide into the tent. Dan took this opportunity to do another assessment, checking over my wounds and dressings and the state of swelling and bruising. My toes still looked as gruesome as ever. The two circular slices across the tips of my first two toes had bled through the bandages, and Dan changed those. My face had swollen, with a black eye and scrape framing my right eye—graphic evidence of the bear's might. The cut on my earlobe was crusted over now, with a subtle stain where the blood had run down my neck. My tailbone ached, and when I sat wrong it bloomed with sharp pain, like pushing on a bruise. Almost any movement, in any position, made me wince. It seemed every part of my body had a direct link to the nerves in my tailbone and, worst of all, to my thigh. The main bite was the same as before, a horrendous puncture wound where my skin was torn and opened to a dark hole. If I pulled at the ragged rim of it, I could see the muscle tissue beneath. The surface of the puncture looked worse than before. The torn flesh where the tooth had pressed before it broke through had stopped actively bleeding, but the bandage was soaked. Underneath, the tissues were black with bruising. The skin looked dead. No blood flowed to this spot anymore. The vessels had been so compressed, they couldn't bring fresh blood flow, fresh oxygen, to the tissue. Ischemia—inadequate blood flow, I would learn—was turning to infarction, no blood flow. The tissue was dying.

The dark patch near the main puncture wasn't the only spot. A few inches toward the outside of my thigh was another dark bruise that looked awful. More hid under my thigh, from the teeth of the bear's lower jaw—the crush wounds. Dan checked them all before covering them again with bandages.

The evacuation algorithm. Over the satellite phone, Dan had conveyed the nature of my injury—a near miss, medically speaking—and what I could and couldn't do. He explained that I couldn't walk, but when asked, said I could likely paddle. I'd been able to hold myself up for buddy carries, after all—a far more strenuous activity than paddling.

With Paul and Aaron on the other end of the line had been a doc from the Twin Cities, who was on his first day volunteering as Camp's health officer. Also on the line was the camp director's wife, whose day job was as an ER nurse. They all listened, hovering around the speakerphone in the small office in the boathouse, making notes, checking coordinates, and plotting distances. They had made calls, alerted the Royal Canadian Mounted Police (RCMP), and checked on the availability of aircraft for an emergency evacuation. They ran into hard facts: the resources weren't there. We were over a hundred miles from the nearest "town" of Baker Lake, and as far north from Winnipeg as Minneapolis, Minnesota, is from Dallas, Texas.

In the event that an emergency evacuation, with its attendant risks, was deemed necessary, we would be chartering a fixed-wing aircraft to fly into our remote location, and we would have to designate a landing zone. So long as I didn't get sick, wasn't showing signs of infection, and could paddle and keep moving toward Baker Lake, we would manage the wound. Camp would contact my parents, too, so they could weigh in on the decision. If they wanted me out, I'd be out.

Dan was under orders from Camp's health officer to monitor me as closely as possible during the night. He was concerned about bleeding both internal and external, shock, infection, rabies, and posttraumatic stress disorder (PTSD). There was also the real possibility of compartment syndrome, a painful and dangerous condition when swelling in the muscle tissues causes pressure that keeps oxygenated blood from reaching nerve and muscle. That would be an immediate danger, whose fix is to make a huge slice, giving the tissue a place to expand and thereby relieving the pressure. That was outside Dan's training, and not something we wanted to try. My risk for developing PTSD would be lower if I stayed with the group. We would monitor for the other complications via thorough checks of my vitals, the wounds, and the extremities below them. Rabies was a serious concern, too—deadly if I had been infected and didn't get the vaccine within the next two weeks.

We had a substantial med kit, which we had raided for dressings. We couldn't use antibiotic salve, because it was a deep puncture—a contraindication for the ointment. We had to wash the wound with water, squirting it deep into the hole—a procedure known as irrigation. We had the standard pain meds ibuprofen and acetaminophen, and the more robust hydrocodone, as well as erythromycin for infection. We had only so much of each, though, and so long as we didn't need to use them, we would hold off.

Dan examined the wounds and surrounding tissue, paying close attention to my feet. They would be the indicators of bigger problems that, unmonitored and unchecked, might take my leg. He looked at my uninjured toes, pressing his finger to the tip of each so the flesh turned white, before releasing and watching the capillaries refill with blood and turn the skin a healthy pink again. Feeling around on my foot, he carefully palpated for an arterial pulse. After slow maneuvering, he found one—weak, as they always are, but definitely there. Blood was getting to my feet, which meant my leg was not swelling in a ring around the vessels and cutting off blood flow to the foot downstream. That was a relief. He put a thermometer in my mouth and let it sit there while he finished examining me, replacing bandages and clothes as he went. My temperature was normal. My bleeding was controlled, and blood seemed to be flowing okay. I was good to go, ready to turn in for a night's sleep. But he would be repeating the check in an hour, and every hour through the night.

The helicopters they didn't have were not being sent, and the planes from far, far away were not being dispatched—at least, not yet. The Femmes were long gone by now, far ahead of us. We were still alone on the tundra. We'd be under only our own power to get out. I wanted to stay on the trip. I was terrified and in pain, but the last thing I wanted to do was leave. I did want to get the hell off that island, but that couldn't happen until the slow sunrise of the next morning.

Dan went out to confer with the rest of the group. The sun had dipped below the horizon, and it was well into twilight. I could just see either side of the tent, and the rolling hills of the island. The musk oxen were off somewhere in the shadows. We could not rest easy tonight.

Dan set up a night watch.

> Patience, he thought. So much of this was patience—waiting and thinking and doing things right. So much of all this, so much of all living was patience and thinking.
>
> **—Gary Paulsen,** *Hatchet*

CHAPTER SEVENTEEN

The Night: Day 29, 23:00

Two guys at a time would stay up through the night. They would sit outside between the tents and the bug tent and wait, bear poppers and mace in hand, watching for musk oxen and grizzly bears in the dark. When Dan told us about the watch, I was terrified. *I can't sit out there,* I thought. *I can't sit out and watch!* It hadn't sunk in yet that I had become exempt from some things out of necessity.

I lay in the tent and listened to Dan's instructions to the others. First watch would be Darin and Jean, to be relieved three hours later by Mike and Auggie. Last shift would be Dan, bringing us through to morning.

I lay there listening to the conversation, waiting uneasily for someone to be back in the tent with me. They were just outside, yet they felt far away. To be alone was distressing, as if my simply being alone would cause everything to happen again. I breathed slowly, focusing on my exhalations, and waited for them to come back into the tent.

Before long, Dan unzipped the door opposite me and climbed in, followed by Mike. They each prepared for bed and slid into their sleeping bags. Dan set the alarm on his watch to ring in one hour, when he would check my vitals, make sure the capillaries in my toes were refilling as they should be, and check the pulse in my foot. Then he would set the alarm again for one hour. It would be a long night for both of us.

"You guys good out there?" Dan called to the first watch.

"Yeah, we're good," they said.

I tried to sleep, but my mind raced. I was uncomfortable and soon felt the need to roll over. I am chiefly a side sleeper, though I first fall asleep on my back, and I'd been lying on my back for a while. I started to turn, and pain shot through my leg. I gasped. After catching my breath, I tried flexing the muscles on the back of my body, to the exclusion of those on the front, including my hip flexor. Carefully, slowly, I arched my back and turned my body, rotating to my uninjured left side. Every few moments, I'd feel my muscle flex, and the resulting jolt of pain. I gasped and winced several times before finally coming to rest on my side.

Before I knew I'd fallen asleep, I woke up to Dan's one-hour alarm. He went through the procedure again, checking my feet and going through my vitals. It went faster this time, and soon I was back in the dark, in my thoughts, and trying to find a comfortable position again.

So the night went. Every hour, Dan's watch chimed and he checked my vitals. In between, I slept as best I could, shifting carefully and slowly. At times, I jerked awake with a gasp after my body had turned in my sleep.

"You okay?" Mike or Dan asked.

"Yeah," I said, eyes wide with surprise as I tried to calm my body and my breathing. "I just moved wrong."

Later, I woke suddenly to yelling and the banging of pots.

"What's that!" I said, terrified it was the bear. "What's out there?"

"Just another musk ox," my tent mates said from the dark.

I wasn't convinced, but then the watchmen confirmed it. Knowing it wasn't a bear, I worked to block out the pots and the whistles and fall back to sleep.

Until the next hour.

After many vitals checks, Dan's alarm chimed again. It was time for another checkup and for his dawn watch. He checked my feet and my pulse again and then stepped out of the tent. Dan would be alone on his watch, staring into the dusky dimness by himself. I admired his courage. I could not imagine being outside the tent, much less alone. Pushing the thought from my mind, I shifted carefully and fell back asleep.

Dan didn't show it, but he was nervous. Of course, he was the leader of the group, and the only reasonable choice to go out alone. Two watchmen could sit back to back and get almost 360 degrees of coverage, along with the emotional support of another human. There was also the intellectual support of looking and pointing and saying, "Hey, you see that?" "Yeah, I see it, too; it's a musk ox." But it was a wide field of view for one set of eyes peering through the murk.

Dan would have to keep his eyes and ears stretched out into the deep twilight for hours. With not much sleep and much stress, he was tired, ragged, fighting to stay awake. He hefted the mace and bear bangers in his hands and settled into the silence.

Movement.

Dan turned and looked beyond the tents to his left, into the dark shadow of rolling hills. It was the slow lumbering of more musk oxen. They were headed toward camp. Dan didn't want to yell at them too urgently and wake everyone up in alarm.

"Musk ox," he said loudly in an even, strong tone, as if into a megaphone, "go away!"

That's when it started to rain.

Really? Dan thought. *It has to rain now?* He sat on the tundra and zipped his jacket and pants, resigned to it. It wasn't a hard rain, but it was persistent—the kind of rain that sneaks up on you and stays for days until you and your gear are drenched through. He hunkered down and waited, yelling to the musk oxen every so often in that clear, slow plod of words, as if the creatures might understand.

"Musk ox, go away!" Dan said slowly. He was miserable. The tundra was turning into a spongy, saturated bed of moss, like a soaked carpet. The incessant patter of rain on his hood and shoulders pulled heat from his body and eventually soaked through the outer layer of rain gear. The inside started to sweat, and a cold clamminess spread from head to toe. He was bone tired, too, and feeling the weight of everything that had happened.

Then, through the rain, the darkness peeled back. The sun began to glow, and the drops shifted from deep black to predawn gray.

Then the sun shifted. Gray turned to deep pink. The rain, the clouds, the ground, Princess Mary Lake, and everything on it was pink. Dan looked up at the sky. It was pink, too. The droplets of rain seemed to glow. Then, stretching from the hills of the island to the deep expanse of water was a huge pink rainbow, the only one he had ever seen.

Dawn was near.

PART III

The bear did just that. She drank and ate and kissed and licked and fought the river, and came across. But when she got near Kiviuq's side she shook herself to get the water off her fur. Her big stomach burst, and because the water inside was so warm, it turned into mist and fog. And that was the very first fog that ever appeared on earth.

—Inuit folk tale, "The Grizzly Bear and Kiviuq,"
as transcribed by Kira Van Deusen

CHAPTER EIGHTEEN

Learning to Walk: Day 30

After the fitful night's sleep, waking came with a heaviness. I opened my bleary eyes to a pale glow inside the tent and felt grateful that it was finally morning. Dan was squatting in the tent door, back from his watch and ready for my last vitals check before we started the day. I could see behind him that the island was encased in thick fog. After the check, pleased to see that nothing had gotten worse, he left to call Camp and report on how the night had gone and how I was doing.

I lay there for a while, trying to muster the strength either to roll over on my side for another few minutes or to get up. I wasn't sure I could accomplish the latter and was still thinking about what to do next when Dan came back to the tent. "They want me to open it back up," he said.

"What?" I asked, sitting up on my elbows to look at him.

"They want me to open the wound back up and irrigate it thoroughly," he said, not moving.

Images of healed flesh being pulled apart once more flashed in my mind. My wound was not actually healed yet, but in my mind's eye, it was. Opening the hole would be like parting a great red canyon of muscle and skin.

"Oh," I said, "okay."

"And they said it's going to hurt," he said. "I'll go get the stuff ready for it—be back in a couple of minutes." He rose from the door and headed to the bug tent before I took a deep breath and sighed. Mike was still silently packing on the other side of the tent. We looked at each other, neither of us saying anything.

When Dan came back, Mike cleared out. Dan zipped the doors to the tent, and we slid my mat away from the door so he could maneuver around me. Our med kit contained a 60cc syringe, brought solely for this purpose. Until now, it had been stored in a ballooned sterile pack. Comically large and shiny, the syringe looked like something Mickey Mouse might use.

Along with the syringe, Dan had brought a pot of drinking water, antibacterial soap, and dressings. Our health officer had instructed him in how to mix the soap into the irrigation solution to help clean the wound. We removed my pajamas, long underwear, socks, and boxers. Shades of purple surrounded the outside of the dressings like a cloud, extending from a third of the way up my thigh nearly to the hip joint and across the front. Even my privates were beginning to show bruising.

The dressing already looked dirty, some blood having soaked into the outer Ace bandage. Dan carefully unwound the elastic, rolling it up as he went. The gauze underneath had wide swatches of red and an ugly rusty-yellow stain that bloomed out from the blood. He carefully peeled it back to reveal the puncture, the lacerations, and the surrounding dark bruises. The wound had dried since I last saw it, and no longer oozed. Now it was sticky. Dan gingerly removed the dressing with his gloved hands and put it in a tiny, flimsy red biohazard bag. The dark red plastic, with its stark black logo, seemed out of place, too clean, though its markings said that it was anything but.

Dan pulled carefully at the edges of the puncture, seeing how far it would open. This wasn't as awful as I had expected. The sensation was dull, almost numb, but uncomfortable. Despite his careful working, the puncture remained shut. We would see what the syringe could do.

Dan leaned back and picked up the syringe. He drew the plunger, filling it with irrigation solution until it met the 60cc line. He carefully placed the plastic tip of the syringe near the opening of the wound and aimed it straight down, following the path the tooth had taken, pointing to the bottom of the deep puncture. He hesitated. "Ready?" he asked.

"I guess so," I said, feeling anything but.

Slowly at first, he pushed the plunger. The cold solution quickly overflowed the wound and spattered out. It felt like the tickle of eyelashes, of butterfly wings. He squirted the rest of the solution in, all at low pressure. It soaked my leg and dripped down into the dimples of my sleeping pad, the newly rehydrated blood staining the pools a burnt red. As he irrigated, I felt increased sensation, like the needles of a waking limb. By the time he'd finished the first syringeful, my leg was awake and stinging.

Dan refilled the syringe, moving more quickly now. He looked at me, and I nodded. He pushed the plunger, harder this time, to clean the wound of clinging detritus and flush it out. Almost instantly, the pressure, the eyelashes and butterfly wings, were gone, replaced by intense, searing pain.

"Whoa!" I gasped, suddenly out of breath. He eased up on the plunger, and the pain ebbed. "Holy shit, that hurts!" I said to Dan, more informing him than complaining.

"We've got to do this," he reminded me.

"I know," I said, still breathless. "It just … surprised me."

This was uncomfortable for him, too, but he steadied the nozzle near the opening again and pressed the plunger. This time, my leg blossomed with hot pain. The edges of my vision darkened, narrowing until I could see only what was right in front of me: the little bit of tent, and Dan hunched over my bloodied leg. I thought I was going to pass out.

We repeated the process again. I tried to contain the pain. I failed. In moments, I was breathing heavily, trying to squirm away, until the pain spilled out of me in groaning, then yelling, then growling, uncontrollably until the plunger stopped. When it did stop, I was flooded with slow waves of relief that left an electric hum in the depths of the wound. Each time I reached baseline and my vision started to widen again, Dan was ready with the syringe for the next round. Again and again he went right back to it, and I had to ask him to wait. I felt bad for him. It takes a great deal of willpower to keep up such necessary torture. Both of us knew that it was necessary, so we set to it again and again.

The thin walls of our MSR Prophet did nothing to muffle the sounds for the rest of the group. They continued breaking camp while listening to my crescendoing cries of pain.

Eventually, Dan pushed the plunger on the syringe one last time and stopped. Gradual relief oozed through me, letting the tension, which had held my body stiff during the irrigation, out through my fingers and toes. It ebbed until all that remained was a general din of discomfort. My leg buzzed with energy, ready to fire off a sharp twinge of pain as soon as I moved my body. I was exhausted—not yet out of the tent to start the day, and already spent, drained by the ordeal.

Dan helped me get dressed, slid my boots onto my feet, and tied the laces. Getting out of the tent required two guys. Once I was upright, I still couldn't hold myself up. When we were finally out of the tent, neither of us was interested in eating our granola, though we both did, knowing we would need the energy.

While we were working and eating, the guys had broken down camp. It was time for the long walk down to the boats, barely visible through the thick fog. I slung one arm over Mike's shoulder and the other over Dan's, just as when we were running from the musk oxen. I tried to help with my left leg, but as soon as I did, the muscles in my right leg engaged, sending horrible shocks through me. It took me only a couple of tries to discard this as an option. Again I was stuck holding myself up on their shoulders, letting them bear my full weight. The boats looked far away through the fog and haze. Mike and Dan moved quickly and precisely. Still, we had to stop several times, all three of us out of breath and sore. As we made our way down, the others ferried our gear to the boats. By the time we finally reached shore, the canoes were packed and the guys were finishing their final sweep of the site. The fog was so thick, we could barely see the spot where the bug tent had stood, and we couldn't even see the ridge—just the green and brown where the ground rose suddenly from the tundra.

The fog shrouded our morning in silence. All of us were exhausted and working quietly. From now on, Dan told me, I would have a dry-foot policy. We would do whatever it took to keep my feet dry, whether that meant pulling the boat farther up onshore or carrying me to the seat. I didn't need further avenues for infection and other nasty things like foot rot added to the list of complications.

With our boat pulled far up onshore, Dan and Mike gently carried me to the seat and lowered me onto it. With my hands on the gunwales, I could help lower myself down. My shoulders ached from the ride down the hill, and I looked forward to working out the tension. A few thousand paddle strokes would massage them back to normal through the hard work they had grown accustomed to. But for now I sat in the boat, on land, hands folded on my lap, waiting for the guys to finish loading their boats, for Dan to get into the stern of this one, and for them to push our heavy bow off this ominous island.

My dry-boot policy had a proximity clause. Dan was to be with me for the rest of the trip. I would paddle with him, and we would stay in the same tent. He would tend to my wounds, tell Camp what was happening with me, and do what Camp told him to do with me.

I wouldn't be able to kneel for white water, and I was gripped with worry about the rapids we had yet to encounter. "Get on your knees!" was practically our mantra for paddling swift water, the golden nugget of knowledge from our training that had kept us upright through hundreds of miles of rapids. Dan could paddle rapids solo, I reminded myself. He could on- and off-side brace the boat from any angle, forward and back ferry on his own, and navigate the river and its rapids without my help, if necessary. But now I was stuck in the seat, unable to kneel in the boat and lower my center of gravity. Even with Dan's paddling abilities, I'd be like an unbalanced mast on a top-heavy craft, and punished with tearing pains anytime I shifted my weight in the wrong direction.

I sat quietly in the canoe, thinking these thoughts and remembering white-water training with Mike. I remembered watching the gunwales go against the horizon and tip, like a spherical compass, until we were thrown out into the rapids. I imagined Dan watching our boat flip on the Kunwak, my stiff body in the front, upright on the web seat, unable to move, to shift, until I was spat out into the river.

We can paddle, I told myself. *I can paddle, and that's what we need to do. If there's something we can't safely paddle, we'll get out and walk, and we'll figure that out, too. Whatever comes our way, we'll figure it out. We have to prepare ourselves the best we can and then confront everything as it comes.*

The guys moved around me, finishing loading the boats. They set our heavy packs into compartments and sealed and lowered our food barrels into place so they wouldn't roll around. They determined the correct paddling arrangements and set paddles and water bottles and other personal items in their proper places. The pile of gear onshore steadily disappeared until everything was in our three small boats.

I compartmentalized my thoughts and pushed the gloom from my mind. We were about to escape the island. The rolling slope of the tundra was obscured by the fog. All around me, it hung like a heavy shroud. I could feel it in my breath, too—a heaviness. I wanted off the island *now.* I felt that as soon as we left, I would be instantly safer. The bear was still there, somewhere. I imagined it appearing from the mist and pounding down the slope. I did not want to see the bear again. I couldn't fight it again.

Behind me, behind our canoes, were the open, calm waters of the lake, also completely shrouded in fog. We'd looked out at Princess Mary for a day and a half, and we knew its shape well. We would pull out from shore and turn left, southeast, to catch the pull of the Kunwak down toward Thirty Mile Lake. It could be foggy the entire day, which would make navigating tough, but we'd figure it out.

"You all set?" Dan asked, standing near the front of the boat, breaking my train of thought.

"Yeah," I said.

He walked past me and splashed into the water in his high boots. The rest of the guys surrounded the front of our boat and grabbed on to the gunwales around me. Together, they lifted the bow and scraped it backward into the lake, red Royalex peeling off in thin pigtails against the stone.

The lake was still, and ripples spread outward from our canoe. We waited just offshore for the others to join us. Behind them, the island quickly faded into the mist, and I could no longer see our campsite or the start of the ridge behind it. I knew it would stay in my mind like a shadow, a presence. To be physically off it was a relief, but we still had four guys onshore. Again I imagined the bear charging down from the murky distance.

The second boat pushed off into the lake. *One more boat,* I thought. *Please, before something else happens.* The last sternman found his seat, and the bowman lifted his end and slid the canoe into the water. Amid motionless banks of fog, he carefully stepped into the boat and pushed off, drifting away from the island. I felt a wave of relief. We all were on the lake. The island seemed empty now, as seemingly empty as when we arrived. I let out a breath I hadn't realized I was holding.

The third boat joined our silent flotilla, and we turned away from the island and toward the Kunwak; toward Thirty Mile Lake; toward the girls, Baker Lake, and help; toward care for my tattered and useless leg. We slipped into the fog, and the island disappeared at our stern, swallowed up in the thick gray.

The water was completely still. Before long, the fog began to clear, and soon the glass of Princess Mary Lake reflected a shard of sun through the haze. Looking up, I saw fingers of mist peeling back to reveal the deep blue of sky behind a light patchwork of clouds.

We left the fog behind us until all around was deep-blue sky. The few clouds above were translucent, with sun glowing through them. The entire lake, the river beyond, and the river before were bathed in sunlight. Everywhere but the island. There, a pall of cloud materialized, dropping down and enveloping it as if it had never existed.

After several miles, we stopped for a break. The lake remained perfectly still. Exhausted, we lay back on our packs in the boats and closed our eyes. I'm not sure how long we slept, but after what felt like a long, necessary nap, we slowly came around. Our three boats had barely drifted in the calm.

Renewed, we paddled silently, putting distance between us and those mysterious shores. The lake was as still as I'd ever seen, sky connecting with water and extending beyond it through reflections for miles. The only sound was the calming rhythm of our paddles as they rose and fell.

Except for the occasional wrong move, paddling was surprisingly comfortable. It felt amazing to be moving. Maintaining my usual paddling posture wasn't a problem at all, as long as I didn't lean back. I was relieved to be contributing again in some way and felt the immense relief of, at least for the moment, not being a burden.

Off the island, though, out from under the fog in clear air, I could see for endless miles, and each boulder, each tuft of scrub grass, looked like a bear. I was terrified, seeing bears everywhere. I looked at a boulder and started, gasping and pointing.

"Is that a bear?" I asked Dan, my pulse suddenly racing.

"No, it's just a rock," Dan said.

I saw the bear boulder loping across the tundra and leaping into the water, swimming fast to our canoe and pulling me in. I tried to clear the thought. *It's a boulder*, I told myself, *not a bear.* I had trouble believing Dan and my own reminder. As soon as I'd convinced myself, I found another boulder, on another patch of shore, that must be a bear and must have the same supernatural swimming powers and malicious intent. A boulder, a musk ox—anything out of place—looked like a bear, and suddenly, magically, it would appear thirty feet from our boat, running across the glass of the lake or rising from it like a violent specter. On this went for the entire length of Princess Mary Lake.

After hours of paddling, we entered the narrows that signified the start of the Kunwak River. There, we noticed a grouping of light-gray stones onshore that looked different, arranged by design. I didn't think this was a bear, but neither was it natural. They were arranged in groups, large individual stones standing upright. It was another Inuit site. We paddled up to it and beached the boats. The guys romped around, exploring the site while trying not to trip on the tussocks, which were everywhere. I was stuck in the boat. Even if I could stretch my legs, I wasn't sure I wanted to, though I did feel cramped from being in the canoe for so long. I visualized standing and stepping out of the boat, imagined how my body would move, how my muscles would flex, and, most importantly, how I would avoid flexing the torn and bruised ones.

Thinking through the motions, I decided I could stand up and could walk. I decided to try. I grabbed both gunwales and lifted my body using only my arms. Careful not to flex any muscles on the front of my body below my chest, I pushed upward until I could extend my uninjured leg and hold myself up on it. At this point, the injured leg dangled in the bottom of the boat. I needed to get it out onto the ground while I had my good leg planted firmly in the canoe. Carefully I flexed the hamstring of the injured one and pulled it back, exaggerating the motion and swinging the leg back and over the gunwale. Oddly, there was very little pain. I planted that foot on the ground and locked my knee before leaning back and pulling my other leg out of the boat. As I lifted the good leg over the gunwales, my balance faltered and my torn right hip flexor tensed. I gasped in pain and nearly fell. The other guys had by now taken note and yelled over to see whether I was okay. I wasn't but said I was. I stood there for a long moment, holding myself up by the canoe, waiting for the pain to ebb, to catch my breath and steady myself.

Steadied, I rose carefully until I was mostly upright, before turning toward the Inuit site and the guys, who watched nervously. I twisted, lifting and turning my right foot away from the canoe before bringing my good leg forward to take a short step. With that foot planted, I swung my right leg forward. I'd taken a short step. I was moving!

My hobble was slow, awkward, and uncomfortable. To balance myself, I held my hands out to my sides, my fingers curled into grotesque claws as if grabbing at invisible rails. Suddenly self-conscious, I tried to drop them to my sides but faltered, and was punished for my mistake by sharp stabs pulsing from my hip. I looked more undead than alive, but at least I could move on my own.

Exacerbating the difficulty of the walk were those tall, reedy tussocks I'd seen from shore. I had to lift my leg over them or swing it around them. Walking the fifty feet to the site, I faltered often, but walking on my own was more important than the pain. When I finally reached the others, I asked Dan if I could have some ibuprofen. He said that he had to ration it for the rest of the group for the rest of the trip, so if I could handle it, the answer was no. I was baffled, but as I thought about it, saving the medicine made sense. I told him I could handle it.

Whenever I started to feel comfortable with how I was moving, I had the tussocks, those damn bundles of grass, to remind me of my injuries. They seemed to leap up, to shift at the last moment and place themselves in front of my feet. I gasped at the stabs of pain when I tripped.

By the time I'd reached the stones, the guys had gone off exploring other interesting things in the distance. From my new, slow perspective, they seemed to dart quickly back and forth like a school of fish.

The ancient camp had several more rock circles, and some more rocks set vertically—standing stones of the tundra. Their stark silhouettes were like shadows of those who had placed them, frozen in time. I tried to imagine their camp now, to visualize tents and kayaks and a cheery fire, but nothing came. *Too tired*, I thought.

I looked away from the guys, away from the standing stones, and focused back on the tussocks—how to avoid them, how to step with my good leg, how to swing my bad one. I was learning how to walk. As hard as it was and as painful as the mistakes were, I was overjoyed. Less than twenty-four hours earlier, a bear had my leg in its mouth, could have destroyed it right there. It could have killed me, of course, but it also could easily have maimed me for life, and yet there I was, walking on my own. Walking hurt, but it felt awesome to be alive. It made me giddy. I walked back toward the boats, wincing every few moments, but with a huge smile on my face.

I got near the water, and Mike pointed his little silver Nikon at me. I went to wave him off, but he snapped the picture. The little shutter inside clicked: an image of a seventeen-year-old, standing on his own on the shore of Princess Mary Lake. I was skinny as hell, with that black eye and scrapes across my cheekbones, dried blood from the cuts on my ear, and a big, unstoppable grin on my face.

I hobbled over to my spot in the front of Dan's boat and grabbed the gunwales to reverse my dismount. I set my boot square in the middle of the boat, then readjusted, carefully lifting my injured right leg, holding it back and taking extra care not to flex the bad muscles. I was about to set the foot down when I twitched my thigh. Stabs of pain shot through me and I landed in the seat, gasping. Dan climbed into the boat behind me; then Darin, Mike, Jean, and Auggie lifted the bow and launched us into the waters of Princess Mary Lake.

We turned from the standing stones and pointed our canoes downriver. It was a quiet narrows, where the shores came together and the water from the lake dropped into the river like dark sculpted glass. The current was swift. Soon we were in a deep channel and couldn't see beyond the riverbanks. Much of the river was unobstructed, free of white water—a welcome change given my inability to kneel for serious rapids. We watched lazily for downstream V's and the exaggerated bulge of large pillows formed by boulders deep beneath the surface, far below the draw of our boats.

We paddled down several miles of fast current in the beautiful sunshine. It was easy going, and we made good time. It was almost relaxing. The more time we spent in the river, the more comfortable I felt in the canoe with my injury. I was still worried for when we'd get to a real set, but the moderate stuff was easy. Floating down the simple rapids, we could pay more attention to the dive-bombing seagulls than to the few rocks that dotted the river. Dan initially talked a lot when we would come up to features, but soon we were in a groove with the river and my abilities, and we mostly switched back to instinct and experience.

We skirted some straightforward features, and I marveled at the sun glinting off the clear water as it surged over the rocks below. The bottom of the river, though often deep, was visible through the water and shimmered with dark browns and greens. In the swift current, it looked like a gravel road going by.

At seven thirty that evening, we pulled into camp on the shores of the Kunwak. It had been twenty-four hours since the bear attack, and we had put twenty-seven miles between ourselves and the island. And we still hadn't reached Thirty Mile Lake.

As everyone else bustled about with tents and packs, I hobbled over from shore to the tent area, and Dan stepped away to call Camp again. This was his third or fourth check-in of the day, and his calls were becoming routine. The guys made short work of the tents and the bug tent, and I helped where I could. Once the bug tent was set up, I found a good roost inside and stayed there. I was tired.

After a few minutes, Dan came back and reported on the call. Paul had reached my parents. He had given them the news, and they had the option to pull me off the trip right then and there and call in that plane. But they deferred to me, Dan, and the network of experts that Camp had been in contact with. They were "overwhelmed but supportive" when they heard about what happened, and would get behind whatever we chose to do.

My parents ... The thought of them was abstract. The idea of them talking with Paul—of Dan having just talked with Paul, for that matter—was disorienting. I almost swooned thinking again of the ripple effect of what had happened, the people involved, how it affected my family. I pushed the thought aside. There would be a time for thinking about everyone else, but now we needed to make dinner.

This was about all I could help with, so I chopped heaps of vegetables for our stew. Once they'd splashed into the pot, I was left with my hands folded in my lap. I felt bad that I couldn't help with anything else, that I was slacking off. How could I be so okay and yet so injured? What could I do, and what couldn't I do? Were we supposed to talk about it, or pretend it didn't happen? I think the latter happened most. I was wrung taut, stressed from the emotional and physical injury, with constant pain that was exhausting, and the horror of seeing bears in every rock and ball of grass was degrading. It wore on me.

I was sometimes overwhelmed by this uncontrollable push for self-preservation, even when the stimulus was imagined. That was one reason I was still on the trip and not on a plane or helicopter heading for some hospital. The doctors at home were worried about my leg and about rabies, but also about PTSD. The longer I stayed with the guys, Dan summarized, the less likely I'd get it. So long as my leg wasn't getting worse, I would stay. I was still with the group, still with the people who had been there when it happened. They would understand it best, yet they still didn't get it. Nothing could completely prepare me for what lay ahead. The best I could do was to stay with the group while taking care that my leg didn't fester.

My heightened startle response had calmed down slightly during the day. I was still nervous at the campsite, but only when I was alone, so I worked at not being alone. In the bug tent, we were all together, and I felt safe. On the tundra, I was always with someone, and this helped make it tolerable when I got startled—when something, real or imagined, loomed out from my periphery, and my heart thumped in my ears. In the tents, our tiny tan domes, I felt safe. The thin nylon, though offering no actual protection from marauding beasts, provided psychological protection. It was home no matter where we were, and with the door shut, the tent felt impregnable. I liked that.

The tents were also where Dan and I would go to do my medical exams and procedures. He'd been told on the phone to irrigate the wound again. I remembered the terrible ordeal from the morning, the cold water like molten stone in my thigh, feeling as if it were spreading through my core and out of my head. I was hesitant and apprehensive. But if it was something that needed to be done, we would do it. We would figure it out, and I would endure it.

Alone in the small space, we uncovered my wound. The bandage had rolled from a wide band into a narrow strip, just covering the main puncture and the lacerations around it. My leg felt warm. My body knew that something was wrong, and was working to fix it. Dan pulled off the Ace bandage and the gauze and put more material in the little biohazard bag. My leg was stained with dried blood, and the puncture looked darker and smaller than before. It hadn't closed or really changed. It looked more as if the tissues around it had swollen and shrunken the opening. Dan probed at it carefully before preparing the syringe.

"You ready?" he asked, his face shadowed behind the glow of his headlamp.

"I guess," I said. How could I ever be ready for this? I'd experienced it once before, but I didn't think I could be truly ready for a repeat without heavy meds.

Dan placed the tip of the syringe next to the opening in my thigh and pressed the plunger. Water spurted into the puncture and splashed out of my leg. He started with low pressure, gradually bearing down harder with his thumb. I felt the tissues in the puncture part against the water. It was a distant sensation. This was a relief. I'd been expecting the unending build of burning pain.

The relief didn't last long. The distant feeling persisted until the water broke through the swollen opening and made its way to the nerve endings again. They cried out, shocking my entire nervous system with loud electricity. I gasped at the pain.

Though intense, it was only half as bad as it had been in the morning. This was a relief compared to the earlier irrigation. As Dan kept flushing water through, the feeling became more distant still. Perhaps I was getting used to it, or the water was numbing the nerves. Either way, this was much better than what I'd gone through twelve hours earlier. It was still awful, just much less awful.

Dan finished irrigating and carefully dried my leg before dressing it with fresh bandages. We had only so much gauze, so we were carefully rationing it, too. This also meant we were careful not to throw away something that could be reused. Not the best practice, but we did what we had to do. The Ace bandage wasn't too bloody, so Dan wrapped it around my leg and the new dressing and secured it with the little metal clips. Instead of slipping back into trail clothes, I slipped into my fleece pajama pants and readied for bed. I was tired, and not the kind of tired I'd grown accustomed to over the previous twenty-eight days. This was not exhaustion of arms and back and legs. This was a new kind of tired. On top of the muscle fatigue was a beaten-down feeling. I felt drained, as if the very air had been forced from my lungs. The feeling pulled at a deep memory from when I was a child on the playground. There was blue sky, motionless white clouds, and the struggle for air. I had launched from a swing and landed flat on my back, wind knocked out of me and lungs unable to pull in air. It had been fleeting, though, a moment of empty gasps. I felt it again now: that full-body ache, the feeling that if I opened my mouth, air might not enter my lungs. I steadied myself, letting out as slow a breath as I could manage. *You're okay*, I thought.

I looked at the sun, now below the horizon, before turning to the tent. Slowly I made the few steps to the door and eased myself to the ground to take off my boots. I scooted my butt into the tent with my arms before pulling my maimed leg in through the opening. I shut the door and prepared for sleep.

So far, I had lived one extra day.

> The day would be the repeated pattern of the hour; the week, the repeated pattern of the day; and one would scarcely be distinguishable from the other, even as an interval in time.
>
> **—Admiral Richard Byrd,** *Alone*

CHAPTER NINETEEN

Loss: Days 31–32

I slept for twelve solid hours before I began floating back up into consciousness. From a dark void I became self-aware and could feel my body. My eyes stayed shut, but I sensed my deep, regular breaths as my lungs filled slowly and exhaled easily. I felt the weight of my limbs and the warmth of my sleeping bag. Slowly I started piecing together my surroundings. The person next to me moved, and I heard the swish of his sleeping bag. As the sphere of understanding and placement around me grew, I realized I knew which direction I was facing, and then I knew where the tent was and its orientation. My mental map was placed and set. I knew where I was in the tent, where the tent was on our site, where our site was on the river, and where the river bent in the vast tundra of Nunavut.

Feeling grounded and placed, I opened my eyes to the pale khaki of the tent roof. I looked over to the rustling, surprised to see someone I didn't expect. I turned toward the tent door, whose nylon was peeled back so that all that stood between me and the vast tundra was the thin mesh of the window. I realized that this, too—the door, the space beyond—had been misplaced in my mind. Like a huge whirling set of concentric disks, my mental map spun from where I'd thought it to be to where it truly was. I suddenly felt a sense of stillness, of motionlessness in an ever-spinning world. I lifted my head and felt the tug at the wound in my thigh. It wasn't just a dream.

The nylon fabric of the tent whipped, and I sensed the vastness of the tundra, felt the drive to reach Baker Lake. *Slow and steady*, I thought. We couldn't afford another mishap. Careful to avoid moving or bumping my leg, I shed my sleeping bag and switched from sleeping clothes to trail clothes, the same I'd worn the day before—the set without the holes and blood.

Gradually, everyone left the tents with their individual gear all packed, and started to strike camp. Dan was working on breakfast: a delicious concoction we called eggs-po-cheese. I joined him in the bug tent. The hash browns were already hydrating in a pot of water. He was mixing the powdered eggs to make a bright yellow paste. With every bit of powder saturated, he set the bag aside. Once the hash browns were plumped out, he drained them and put them in the skillet. They sizzled in the hot oil. Dan let them sit before turning and stirring them. Their bottoms were a toasted gold, steaming in the sun. When the hash browns had all nearly cooked, he poured in the egg, stirring it quickly. Just like real eggs, they cooked fast, and he added the cheese. The egg stuck to the pan, and he scraped at it. In a moment, he lifted the pan off the stove and set it on the ground. It was a gooey masterpiece of melted cheese and not-yet-overcooked eggs. He divided it into our six cups, the cheese stringing from the pan with each scoop. The potatoes were like little french fries. This was one of my favorite trail breakfasts.

With food in our bellies, Dan and I retreated to the remaining tent to look over my leg and check the bandages. He shut the doors, and I uncovered my leg. The Ace bandage was in nearly the same condition as the night before, and so was my leg. Forty-two hours can make a big difference, though. The wound looked older. Parts of my leg below the skin surface were visible through the torn flesh and looked like meat that had been out too long. The edges of the wound had dried, like the rims of a dark, rusty canyon. There were no red flags but no real signs of progress, either. Dan covered it and wrapped it again, readying it for the day as best he could. Luckily for me, it was almost completely numb by now, and the light touch, cleaning, and bandage application felt like distant prodding of a sleeping limb. Every once in a while, I felt a shock of pain, but mostly I felt the numb tingling, which I much preferred over the stabbing pains from when the wound was fresh.

After the bandage change, I got dressed and threw the last of my things together before clambering awkwardly out of the tent. I still couldn't tie my boots properly, not my right boot, anyway, and Dan secured it to my foot for me. With help, I stood and hobbled away from the tent, toward the boats. The guys had taken down the bug tent and packed away all the kitchen stuff.

In the time it took me to hobble to the canoes, Darin and Auggie took down and packed the tent I'd just vacated. I shoved my things into the waiting canoe pack and did what I could to cinch it shut. My movements were slow and at half power. By the time I was standing next to the boat and ready to climb in, everything was gone from camp, and all the canoes were loaded and ready to go. The guys pushed us off and climbed into their boats, and we left to start the day. I looked at my watch. It was two thirty in the afternoon.

The river was dark and reflected the gray of the sky like obsidian, glassy and sharp. It again pulled us at a fast clip. I still felt vulnerable sitting tall in the bow. We hadn't encountered any real white water, but I felt like an upside-down pendulum. *We'll be fine*, I told myself. *We'll be fine, or we'll pull off the river and walk it; I can do that now.*

Dan and I talked over the plans for the rest of the trip while we paddled. We'd keep going, reach Thirty Mile Lake, and keep pushing forward to catch the girls. Once we reached them, we'd have a cache of fresh dressings from their med kit, extra unused batteries from their satellite phone, and support from people we knew and trusted. That would help us through the last string of lakes and river to Baker Lake. So long as my wound didn't get infected and we reached that little town, I'd be fine. We just had to reach the girls first. We didn't know where they were or how far ahead of us they might be, but we figured they were at most a day ahead. But we could put only so much more demand on ourselves.

Every so often I still moved wrong, flexing the front of my right thigh. It felt like a hot brand in the wound, and the burn emanated in a surge of energy that jolted my muscles like a heavy pulse from an electric fence. I did my best to avoid any movements that caused the sensation. At times it was so overpowering I was worried I'd drop my paddle or throw myself so off balance that I would fall back off my seat or tip out of the boat itself. Mostly, though, I felt strong in the bow. I couldn't paddle as hard as before, because real paddling is a full-body endeavor, but I felt the hard pull of the water, and the muscles in my back and arms as they worked against it. It was wonderful to be able to contribute to the group, to contribute to my own rescue.

If we somehow missed the girls, our next supply depot would be Baker Lake. We were headed there as fast as we could. In that tiny hamlet of seventeen-hundred souls, there would be a medical center with supplies and equipment, staffed by professionals. In their cooler would be a vial of lifesaving rabies postexposure vaccine. Without it, I could die in convulsions and delirium. The Camp health officer had spoken with experts from the Infectious Diseases Clinic, the Centers for Disease Control, and their Canadian counterparts about rabies and my case. According to them, I had twelve more days to reach the clinic and get the first injection of the series.

Before Baker Lake, we would have to traverse Kazan Falls. Known for their beauty and ruggedness, the falls were always going to be one of our biggest portages, but now this would be even more challenging. We would reach the portage, and the guys would carry all the gear while I slowly worked my way down the trail, nearly empty-handed. Maybe I could carry my daypack, I thought. It would be slow, and painful at times, but it would be fine. From there, we would skirt some big white water and portage more than we would paddle, before being spit out into Baker Lake and the huge delta of the Kazan River. But all that was thousands of paddle strokes away. We still hadn't gotten to the start of Thirty Mile Lake, and we had seen nothing of the Femmes. All those carefully laid plans were away in the future, blurry forms of what was to come.

We'd been warned about tricky white water where the Kunwak and the Kazan rivers met west of Thirty Mile Lake, and we were ready for it. As we floated down, we waited for pillows to bloom up from the calm river or for the water to drop out in a hidden ledge, but it didn't. The river was calm, a swift but gentle flow. The Kunwak spilling into the Kazan was like a capillary meeting an artery.

Once in the Kazan, we drifted easily. The river was so wide that, but for the pull of current, it could have been a lake. The water was flowing swiftly, though it was so gentle and smooth that we decided to break out our lunch and eat while we floated downstream. We pulled the three boats together and ate quietly, all watching for changes in the river as we drifted toward Thirty Mile Lake.

Just before the river met the lake, we packed our food and broke up our flotilla—and none too soon, because once we were apart, we drifted past a point that formed the western end of Thirty Mile Lake. North and west of the point, an enormous bay stretched out like a huge thumb. Wind whipped across the bay, stirring Thirty Mile Lake into a frenzy and pushing the water so much that it created its own current. The bay met the flow of the Kazan in a fury of waves bouncing up and down against one another with no discernible order. We steeled ourselves for the rough going, and if I'd been able to, I would have dropped to my knees in the boat, like everyone else. We hit the waves and were immediately rocked back and forth, bounced up and down. Dan and I kept paddling, kept our blades in the water as much as we could, balancing the boat with our paddle strokes and the movement of our bodies. Mine felt awkward and stiff and painful, but I was careful not to lock my body.

We kept paddling and finally made it to an island across a small neck of water from where we planned to camp. When we'd last seen the girls, they showed us a map of Thirty Mile Lake, marked with the locations of campsites, graves, and inukshuks. We had photographed the map, and reviewed the image on the one-and-a-half-inch screen of Mike's camera as we were approaching the lake. The island we'd found broke the waves at the far end of the bay, and we took shelter there briefly before paddling the small bit of water to mainland a couple of hundred feet north. Even here, we could feel the pull of the current. It seemed the Kazan pushed the whole of Thirty Mile Lake into a wide expanse of river. There was an energy here, in the water and in the air, almost a presence. It felt as if we were being watched by great players from above.

Having crossed the small channel to the mainland, we beached the boats on the gentle slope of the north shore and began making camp. By this time, the heavy clouds that had been darkening throughout the day began to release their rain. It came down in drizzly sheets, whipped by the sharp wind from across the bay and the tundra.

It was evening, our scheduled check-in time, so Dan took the satellite phone and went to call Camp. It was a long call. They reiterated the importance of keeping the wound clean and getting to the rabies shots in time and informed Dan that I'd probably end up talking to the press at some point.

We tightened our rain gear and set to raising the tents, pitching the rainflies first to protect the inner tent from the thickening storm. By the time we had both tents up, our rain gear was soaked. There were no bugs. We seemed to be the only living things stupid or desperate enough to still be milling about. We opted once more to eat lunch for dinner so we could avoid cooking and go to bed earlier.

Cleaning up after our meal, we storm-proofed for the night and cleaned up the random mess that had formed in our short time there. I helped with the light duty of picking up gear and bags and stowing things where they wouldn't blow away. Everyone milled about, tightening the straps on portage packs, putting away personal stuff, and stowing gear in the boats and the tents. Back and forth the guys went to the boats, filling them, then pulling them farther up on shore. The slope of ground there was so gradual, it seemed we could walk out a quarter mile into the lake before the water would touch the tops of our boots. The guys heaved all three canoes up onshore so that only the last bit of their sterns were licked by the slight rise and fall of waves. The loop-three food barrel, our last, with all our meals for the rest of the trip, whether that was four days or the full week it might take to reach Baker Lake, was standing in the last boat, near the bow. It looked like a smokestack on an old tugboat. This canoe was the farthest north, the most protected, the heaviest of the three. It marked the end of our little harbor.

Having finished putting things away, I limped off to relieve myself and brush my teeth. The sky was dark, mottled ripples of leaden gray, the tundra various shades of shadow and black behind sheets of heavy rain. We were in clouded twilight. When I came back to the tents, most of the group had turned in. I looked around camp to find everything ready for the night, tucked away, packed, sealed, and secure. My eyes scanned for those last few things that might have been forgotten, and found none until I looked at that last boat. The guys had already pulled it up, but some of its stern was still in the water, and from where I stood, it looked as though it should be farther up from the water. Without saying anything, I hobbled over to it.

Turning into the wind let waves of rain into my hood, and I shielded my face. Standing in front of the canoe, I gripped the handle in the bow and pulled up. It was heavy. The last food barrel in the bow was like a heavy nail holding the boat to the ground. There were a couple of packs in the boat, too. One of the large blue portage packs sat awkwardly, packed poorly with everyone's dry boots but mine. Everyone was too tired and hadn't cared enough to bother putting them on this evening. With my dry-boot policy, mine had been on all day.

Far in the bow, almost wedged into it, was my blaze-orange backpack. Near the pack sat the black brick of my Pelican case, with camera inside. There were extra paddles and other odds and ends, and that heavy food barrel. This should be good enough. Still, I grabbed the boat and hefted it another few inches onto shore, as much as I could with my leg protesting, but not as much as I wanted. I was only double-checking, after all. The other guys had already battened down the hatches.

I had managed to pull the other two boats a little bit but could hardly budge this last one. I let go, exhausted by the effort. *Damn I'm weak*, I thought. I took two awkward steps around to the bow compartment and reached in to grab my camera, driven rain stinging my face. I grabbed the handle and hefted it out of the boat, turning back to the tents. Walking here was the easiest it had been since we left Princess Mary Lake, thanks to the smooth ground and my arduous practice in the tussocks. A few awkward steps from shore, though, before I even reached the pile of gear around the wanigan, exhaustion caught up with me. I set the heavy photo brick down in the middle of an empty patch of sponge. Free of the awkward weight, I hobbled to my waiting tent and sleeping bag. It was almost all I could do to get to the door, open it, and shuck off my boots.

Tired, and relieved to be tucked in and dry, I closed my eyes to the tundra, lulled by the white noise of wind and rain.

I gradually woke to the dim glow of the tent cloth, and the sound of wind and rain beating against it. It hadn't stopped the entire night. We had hoped for a break, but it kept coming in a relentless barrage. It was cold, too, in the forties. We would not be moving. For now, we were stuck at our little site on the shore of Thirty Mile Lake.

Jean climbed out from the other tent to relieve himself and see the storm. He walked through the rain and looked toward the open lake, then at our small fleet of boats. There was the one closest to the lake, the second, and then … He looked around, up and down the shore, past the first boat, and back to where he remembered beaching the third the night before.

"One of the boats is gone!" Jean yelled over the wind.

Hearing this, Dan jumped up out of his sleeping bag and threw on his clothes and rain gear. In half a minute, he was outside the tent, quick-timing it to our harbor of canoes. The closest canoe was there. So was the second. But the third was gone, in its place a blank spot of undisturbed tundra. He looked at the other two boats, at the gear stowed in them, and at the pile near the wanigan. Anything else was gone.

The loss of the boat was devastating. With the rain, millions of gallons of water had saturated the tundra. It spilled into the lakes and traveled downstream, filling Yathkyed and Forde Lakes and the Kazan River to the southwest, and Princess Mary Lake and the Kunwak River to the northwest. At the west end of Thirty Mile Lake, just beyond our campsite, these rivers joined. Their influx of rain had raised the water level by nearly a foot. Our boat, whose beaching had been solid the night before, had been surrounded by the rising waters, lifted off the ground, and carried out into the lake.

Along with the boat, we were missing Auggie's paddle, my spare plastic paddle, our big barrel of onions, my fishing rod and reel, my blaze-orange daypack, my gloves, my compass, my bug jacket, Darin's fleece, my fleece hat, a dry bag, the portage pack with everyone's dry boots but mine, several water bottles, and, worst of all, our loop-three food barrel with our sustenance for the rest of the trip. It was the boat I had tried to pull up farther the night before. Our food was gone. We had cocoa, coffee, sugary drink mixes, desserts, brownies, cookie dough, batters. In the wanigan, we still had all our cooking supplies, oils, flour, butter, spices, and breading. The one part of our trail food we needed most was substantial meals. And they were lost.

Both the satellite phone and my camera had been in the boat earlier, but we had taken them out the night before. I remembered struggling with the boat, giving up and coming around the side to grab the Pelican case. I'd felt compelled to move it, wanting to bring it closer to the tents, and there it now sat, barely ten feet from the boat's empty berth.

We had the basics for cooking, but we would have to catch fish for our meals. We'd had success fishing, and with what we had in the wanigan, we could make batter and cook it. We would be eating fish, drinking sugary drinks, and mixing up desserts. That depended on the fish biting. A couple of meals without their protein, and we'd be hurting. It wouldn't take much for us to get behind nutritionally. We had already felt the effects of not having quite enough food, substituting lunches for dinners, fish for meals, and extra desserts for lacking entrées. We'd made it work, but now, instead of being just low on supplies, we were almost out of them.

We scanned the horizon. The unrelenting rain and wind made it hard to peer through squinted eyes and rendered the binoculars all but useless. Our initial search turned up nothing.

Our big, bright-red canoe, the only thing like it in the vastness of water and tundra, was gone. We weren't up shit creek without a paddle—we didn't even have the canoe! Inventories of our gear and our available canoe space spun in my head, parts and packs dropping in and out of the boat as if on an assembly line. The configuration of people and matériel gradually locked in until the two remaining boats were filled with six people, and only a few floating pieces of unplaced gear remained. *Shit, that will be tight*, I thought, *but we can make it work. Tetris. We'll just have to play the old block puzzle game and fill every space.* It was no different from packing a backpack, a camera case, or a car trunk. We just had to look at the pile of gear and the small space it had to occupy, and move things around until it all fit.

After a long period of silence as we each thought through our new predicament, we discussed the options. Two things quickly became clear: we could survive without our food barrel or any of the gear we'd lost, but we had to try to find it. Dan and Auggie would wait for the weather to part, for there to be calm, and then go out in a boat and search. The decision to go out alone and separate the group came after carefully weighing the options. Only after great consideration did Dan decide it was worth the risk. They were some of our strongest, smartest paddlers and could handle a boat in adverse conditions. They would go out with a specific mission, a specific timeline, and the right equipment.

But the thought weighed heavily.

With the dire turn of events, none of us was hungry, so we skipped breakfast. The storm raged on. I stayed in the tent and rested, sleeping most of the day, waking regularly only to look around the small shelter before going back to sleep. The tent felt like a hospital room, stuffy and plain, a place you rarely left. In between periods of heavy sleep, I listened to the wind and the rain. I don't think I dreamed.

Throughout the day, my naps shortened until I was waking often. The tent got stuffier and stuffier. I shed layers and then my sleeping bag. The storm never abated.

Hours into waiting, we ate a sad and soggy lunch. Dan looked at the clouds and scanned the horizon. With afternoon came the break they'd been waiting for, and it looked as though the weather would hold for at least a while.

Dan turned to Auggie. "How you feelin' about paddling?" he said.

Auggie looked back at him. "Let's do it."

With the wind whipping and sheets of fine mist blowing across the vast openness, it was still a gale, but nothing like the tempest we'd sat through since arriving at this place, and certainly nothing like the storm that had taken our boat. This would be manageable.

Dan and Auggie began preparing for their mission. I wanted to volunteer, but something told me I wasn't eligible.

I felt at least as responsible as the rest of the group that the boat was gone. I had been the last one to touch it, after all. *I should have yelled to one of the guys to come out and haul the boat in,* I thought. I was weak, injured, and, though I didn't know it yet, sick. I had tried to help haul the boat up more, but it hadn't been enough, and now I wished I could go out with them—not to redeem myself, but because I wanted to help, to do what needed to be done. Instead, I stood and watched Dan and Auggie get ready, my leg pointed forward to ease the stretch on the wounds.

It was three in the afternoon. Dan prepared us for their absence, going over the plan several times. He left the satellite phone with us, confirmed that we knew how to operate it, and told us to call Camp at nine that evening if they weren't back. I didn't even consider the possibility of their not returning by that time, but the burden of it weighed on me. We all optimistically pictured the missing boat just around the bend, just around the corner.

I remembered the lost boat on my Norwester trip from the year before, how we'd found it just a quarter-mile away. I probed at that memory out of hope. We'd been lucky. But that lake was much smaller, it wasn't part of a river, and our food wasn't in it. And there had been trees. Here, there was nothing to keep the boat from rolling like a tumbleweed. I pictured the map in my head, remembered the current and wind, and thought of the huge span of the aptly named Thirty Mile Lake. It could be anywhere.

It could have been floating for over twelve hours by this time. Or it could have sunk.

Over the course of the day, the wind had swung from northwesterly to westerly. We suspected that our canoe had stayed in the lake, floating south with the first winds before being blown east down Thirty Mile Lake as the gale shifted.

Dan and Auggie cinched their rain jackets, clicked the buckles on their vests, and climbed into one of our two remaining canoes. We wished them luck, and they pushed off from shore in an empty boat. They turned and paddled hard, straight into the wind. In a moment, they were in open water, rising and falling as they crested and then disappeared behind wave after wave. The weather was still not good, and they needed to get back before it got worse. They paddled with purpose, turning nearly broadside to the waves to cross the lake and make their way to the south shore.

In all directions, the horizon was dark gray, like the cold, rough hull of a battleship. But over the lake, our site, and the canoe out in the water shone a ray of hope. Golden sunshine hit their boat and sparkled off their paddles. It felt as if we were in the eye of a hurricane, though, and I wondered how long it would be before the fury of wind and rain pummeled us once again. I hoped Dan and Auggie would be back before then. I watched them shrink into the distance until they slipped behind the massive island to the south.

We tried not to watch the clock. There was no point; they would be gone until they came back. I stayed in the tent and did nothing but lie there, sleeping and waking. Several hours later, I took to sitting. My back would let me lie down only so long, and I'd been doing it most of the day.

At just after eight thirty, I looked at my watch. It was half an hour from call-Camp time. There was still no sign of them, and the weather had deteriorated. I sat in the tent, my nose against the mesh of the door as I watched the open water, the sharp rolling waves. The lake looked the same as when they'd left, though it had darkened and the waves had grown. *I don't want to call Camp*, I thought. *I don't want to have to.*

I watched, and the minutes clicked by. At eight forty-five, after being in the tent nearly the entire day, I noticed through the scrim of the mesh and the portal of the tent vestibule an anomaly on the water. I squinted out. The image kept appearing and disappearing, like the flicker of a mirage. There it was again—color amid the gray. I suddenly wished I'd kept the binoculars. The next time the mirage appeared, it stayed. I unzipped the door and looked to the distance. Free from the blur of the mesh, I could see now. It was the bobbing red form of Dan and Auggie's canoe, on a course straight toward camp. Relief swelled in me.

They bobbed up and down on what seemed like endless waves, the two of them paddling hard the whole time. Gradually, they grew from the distance, until they were passing between the island and our shore campsite. As they approached, it became apparent they weren't towing a canoe. Closer still, the boat looked empty. We were relieved to see them but deflated that our important supplies were still lost. Seeing them again was far more important than the equipment. I had hated the thought of calling Camp to report that not only had we lost a boat, but our guide and one other camper had gone looking for it and were overdue on their return. We were spared that, at least. The four of us waiting at camp crowded around the landing.

When Dan and Auggie were within shouting distance, Dan shouted, "We found the food barrel!"

"What!" we yelled back in disbelief.

"Not only that," Dan said, clearly exhausted but nearly giddy, "it's dry on the inside!"

They dismounted, walked the last few steps to land, and heaved the boat far up onto the grass. There was the barrel, gleaming bright blue. We were elated. The barrel was beastly heavy, and all the scenarios in my head had the canoe tossing it and the rest of its gear out into the middle of the lake. From there, it was sink or swim, and there was no way in my mind the barrel would swim. But somehow, it had.

Dan and Auggie stretched, beat from their ordeal but energized to be back at camp with their find. Along with the barrel, they'd found my tube of sunscreen, the onion barrel, and Auggie's Crazy Creek chair. We pulled the gear out of the boat and put it away.

Both Dan and Auggie were soaked up to the waist, despite their rain gear. We'd been in radio silence since they disappeared at the horizon all those hours ago, and they started to tell us of their little adventure.

After losing our campsite off their stern, they had continued across the lake in heavy waves. As the far shore had slowly come into view, so, too, had a great blue beacon: the food barrel. It was on its side on a smooth beach, partly buried in the sand. Invigorated, Dan and Auggie paddled straight for it. As they neared shore, Dan, overcome with joy, leaped out of the boat, splashing the last few feet to the barrel before bending down to kiss it. He was elated. Heaving it upright and out of the sand, he found the lid still on. He wiped sand from the top to reveal the aluminum lever that held the lid and the seal. Inside, he knew, would be a slurry. Some of the bags would have leaked, letting water in and food out; the barrel would be a tea of broth and spices.

As always, the lever on the retention ring was stiff, and he worked at it awkwardly with his cold, wet hands. It finally popped open, and he set the ring aside. The lid was now held on by a tight fit in a small groove, like a huge Tupperware container. When he peeled it back, no water came splashing out.

Setting the lid aside, he explored the first layer of meals. The top bags were intact, even dry. He reached in and rifled through the rest. There was no water in the barrel. The seal had held. We wouldn't have to go from meal to meal, hoping the fish would bite, sustaining ourselves on an unreliable protein supply and drinks and desserts.

Dan laughed, dropped to his knees, and embraced the open barrel, kissing its sandy side once more.

"It's all dry!" he yelled to Auggie, who was standing in the water, holding the canoe as it bounced and crashed in the waves.

Dan popped the lid back on and, with the barrel sealed, hefted it over to the bouncing boat.

They were far from camp and would have to beat upwind all the way back. Dan reviewed the map, tracing the shore to find a section of the lake in the lee of the wind. Not far away was a point they could hide behind. There they could walk the boat alongshore, but without portage straps for the barrel or a tow strap for the canoe, they would have to stay in the water. Paddling close to shore in heavy waves and wind is tough, so they walked in the water, as if walking the boat through a set of white water. Unlike the shore on the north side of the lake, the beaches of the south were steep, and Dan and Auggie had trudged through deep water much of the time, fighting with the boat as it rose and fell several feet with each wave, the crests breaking up to their thighs. Had it not been for their exposure to cold over the past month, and the hard physical work of hauling the boat, they surely would have been hypothermic.

As they walked along, they came upon the tube of sunscreen, the Crazy Creek chair, and the onion barrel.

Finally, Dan and Auggie reached the spit of land that would take them across Thirty Mile Lake back to our site. They couldn't see us yet, but using a set of islands as stepping stones, they would get back. For the bigger expanses, they would shoot a sight line to a distant island or grab a bearing with map and compass.

With their navigation dialed in and the next wave of storm threatening in the distance, they pushed off from shore, into the tempest once more. They got into the boat as if starting a rough set of rapids, dropping to their knees and paddling hard. Soon they were bobbing up and down over sharp waves on their way back to camp. Dan had looked at his watch. He knew they would barely make it back before nine o'clock, zero hour, when the rest of us at the campsite would have to call Camp and inform them that Dan and Auggie were overdue.

They paddled furiously against the wind and the waves. The light canoe, with only their skinny bodies and the relatively light load of the rolling food barrel, was more like a kite than a boat.

Eventually, they rounded a spit of land and saw the pale domes of the tents, and the solitary shape of the one remaining canoe. They paddled hard until their boat slid up onto the wet grass.

Just minutes before the deadline, they had returned to a cold, wet, windy camp. They'd been gone six hours. Dan and Auggie were chilled and drenched, but thrilled to have found our food. We breathed a sigh of relief and stood rapt as they told us.

I imagined my foolish daydream of going with them, my nearly useless leg, soaked to the nerves from walking in the water, the mess of my dressings, and the resulting infection. If I had been with them, we'd still be out there.

While my thoughts wandered, we prepared for dinner. We popped open the barrel. From the wealth of food inside, one of the guys grabbed a bag of the rich peanut *gado gado* sauce we had prepared at Menogyn. We'd mixed peanut butter, soy sauce, and all manner of spices and tied it shut, quadruple-bagging it to prevent leaks. We boiled shells and mixed them with the rich sauce over the stove. It made a thick, zesty, peanut-buttery pasta. It was delicious and warm, thick saucy pasta spreading the heat like a happy hearth. After being without our food and staring at the bleak prospect of finishing the trip without it, this simple meal felt like a feast.

Today, we had truly dodged a bullet, partly by luck and partly by perseverance. It hadn't been easy for Auggie and Dan to go out on the stormy lake or for us to stay behind and wonder. We were relieved to have our food again, relieved to have them back.

To celebrate, we passed around colorful saltwater taffies. Giddily we looked to one another and peeled the waxed-paper wrappings off the taffy. The bright pastel colors reminded me of Easter eggs, and I salivated at the sweetness.

After putting all the wrappers in the food barrel, we returned to the tent to check my leg and turn in for the night. Dan unwound the bandage and pulled back the dressing. The torn flesh looked stuck, neither healing nor getting worse. Feeling was almost completely gone in an area that started just above the wound and stretched far below the most distant tooth mark. Instead of the sharp pain I had gotten accustomed to, I felt a warm hum, and all touch seemed distant. Still, it didn't seem worse. Dan covered the wound, and we prepared to sleep.

By the time we were closing our eyes on the day, it had been raining twenty-eight hours straight in the "desert" of tundra where we were told it never rains. The wind had been blowing even longer. We'd also been told that everyone gets stuck on this lake. Everyone gets windbound. Something always happens to canoeists on Thirty Mile Lake. On the map, it looks nonthreatening—a thin, long lake running almost due west to east, dotted with islands. The islands are home to Inuit graves and enormous inukshuks. The shore is mostly flat, the lake so narrow it's almost a river. At least three waterways spill into it, and it marks the last big lake, the last big body of water before the Kazan shoots down some forty miles of winding river, over rough white water and down Kazan Falls before miles of bubbling rapids and, finally, its wide delta at Baker Lake. It felt like a checkpoint, as if we were being reviewed by a gatekeeper who would decide whether to let us pass or to hold on to us.

I pushed the lake from my mind and once more closed my eyes to sleep under the constant whoosh of wind and rain.

> And only the enlightened can recall their former lives; for the rest of us, the memories of past existences are but glints of light, twinges of longing, passing shadows, disturbingly familiar, that are gone before they can be grasped.
>
> **—Peter Matthiessen,** *The Snow Leopard*

CHAPTER TWENTY

Fear: Day 33

As I lay asleep in the tent that night, I dreamed I was in a cafeteria. In my hands was a plain tan tray, and before me a wonderful spread of brightly colored food. I had no idea what any of it was, but it looked like all sorts of frosted delights. I scooted down the line and grabbed five of the huge, puffed, sugary wonders and put them on my tray. They all were different shades of blue: bright blue like the clear skies we hadn't seen the past two days, deep blue like a painting of the ocean, pale turquoise, and shimmering blues like a silk sari. In my dream, I took my tray of delectable treats away from the grand buffet. Smoothly, as if walking through a doorway, I stepped out of the cafeteria with my tray and onto the spongy grass of our campsite. I was uninjured. I walked over to our tents and crouched through the door, careful not to tip my tray and lose my treats.

Inside the tent, I went on my knees to the far edge and found my sleeping pad. I sat down with my legs outstretched against the side of the tent, the tray on my lap, and started to devour the bright blue treats, like a child on Halloween night.

After a few minutes, there came a rustling from outside the tent. The distinct sounds of something walking, heavy breathing, and sniffing came through the tent walls. I knew those sounds. I could hear and feel the weight of the paws on the ground. The bear had come back, drawn by my tray of blue sweets. It kept sniffing with short, searching breaths. It had followed the scent until it was just outside the tent wall. It could sense the huge cache of food just inches away.

I sat frozen, watching as the bear pushed its nose against the tent wall. It swung left and right, searching, making its way to where I sat. I could see the subtle shake of the nose as it inhaled, still pressing the tent. I sat still, waiting for it to leave, my body tense as I held my breath. Its nose wandered down toward the tray on my lap. It traced the line of my femur to my knee. It stopped suddenly there, as if it had found its goal. It waited. Then its nose, pushing against the tent like a macabre puppet, shook with two sharp sniffs. It held still after those breaths, and after a long, motionless pause, it pushed its head forward. The shape of jaws widened on the tent fabric.

The bear spread its teeth and, in one swift motion, put them around my knee. As soon as they were in place, it clamped down. I felt the sharp pain of punctures and the terrible pressure. I yelped and grabbed at the bear's head. It had my knee firmly in its jaws. The pain was too real, and I could feel everything—the damage to my knee, the hardness of its teeth on my fingers, and the soft, smelly gums and callused lips—as I tried to part its jaws.

I yelled out. Now it was pulling at me, trying to tear me through the tent. I latched on to the bear, reaching over and grabbing the fur of its jaws. It held on tight. I heaved, and grunted, pulling at the jaws to relieve some of the pressure on my knee. I gritted my teeth and put every ounce of my strength into prying them open. Slowly, they started to part. The teeth pushed against the flesh of my hands and felt as if they would tear through at any moment, but that didn't matter. Gradually, like the parting of enormous ice floes, I pried the jaws off my knee. They opened slowly as I pulled on them, until finally I could yank my knee free.

At that instant, I woke up. My eyes flashed open, and I sat up suddenly. My chest heaved as I tried to catch my breath. My knee was intact. The tent was intact. It was dark, and outside, the tempest of the storm raged on. I tried to force my eyes to see something in the dark, and slowly the forms of the rest of the tent and the guys around me materialized. Gone was the tray of sweets. Gone was the bear. My heavy breathing remained, and I felt my skin slick with sweat. My knee seemed to ache still from the dream bite. It had felt that real. I had felt every bit of the bite and every bit of the fight.

Lying there, I couldn't shake the image of the bear's head and jaws pushing through the nylon mesh. No one had awoken to my start, and it felt as if I were lying alone in the tent. I lay in the dark and in my thoughts as the sweat formed in tiny beads across my body. I tried taking in slow, deep breaths. *It was a dream*, I reminded myself. I listened as hard as I could to the world outside the tent. In my mind, I could feel a bear waiting outside, sniffing for those blue treats. But my ears, thumping with my slowing heartbeat, only heard the whoosh of rain and wind.

I tried to sleep, tried to push the dream out of my mind, but it compelled me to be vigilant, to listen. No sounds changed. There was no rustling, no heavy steps, no sniffing. Eventually, the slow breathing calmed my body. It was a long time before I could fall back asleep. I was terrified that the dream would return.

The next morning, we woke gradually. The abnormal storm was still drenching the tundra, and the wind still whipped Thirty Mile Lake into a tumultuous gray sea of waves. We were still bound to our site, captive to wind and rain.

Mike unzipped his door and poked his head out through the nylon. He was surprised to see hundreds of tall, lean, brown animals with huge racks of antlers. It was a herd of caribou, some within twenty yards of the tents. Several turned their heads and gazed in his direction. Their eyesight is terrible, and they probably just saw two dome-shaped boulders near the shore.

We stayed in the tent that morning, reading, playing card games, and trying to pass the time. I worked on my water-damaged copy of *The Snow Leopard*, trying to muster the energy to stick with Peter Matthiessen as he went through his own grand adventure in the Himalaya. The day moved slowly, and we skipped breakfast. We weren't hungry and weren't burning any calories sitting in our stuffy little tents.

Auggie and I didn't leave our tent until three in the afternoon. I hadn't even gotten up to pee when we were finally roused from our nylon cell for a lunch of couscous. The soupy mixture, usually a dinner, replaced both breakfast and lunch, though it was far too late for breakfast and late even for lunch.

Impossible as it seemed, the wind had intensified. Even bringing the spoon to my mouth risked bits of couscous being blown onto the soggy ground. We had to brace ourselves when standing and even sit solidly, or risk being toppled over. Anytime we moved anywhere, wind at our back could throw us on our face, and wind at our face could knock us on our back. We guyed the tents down as best we could—the last thing we needed was a tent blowing into the lake or, worse, becoming a tumbleweed across the endless tundra. For the moment, we didn't have to fight the rain as well, but the gale-force wind was making up for it. Even talking was nearly impossible. The wind seemed to bat our words right out of the air before they could reach the listener's ears.

In the silence of the group and the din of the wind, my mind raced. I found it wandering and lingering on things without my direction. Most predictably, it kept coming back to my leg. Under my rain pants and the green zip-offs I'd switched to after the attack were the wraps of Ace bandage around my thigh. Underneath, my leg felt warm, as if drawing heat from coals. It was almost pleasant in the wet, cold wind. But beneath the warmth was an ache. For the past day or so, it had been numb, and the pain was a novel sensation. While switching from pajamas to day clothes, and changing the dressing earlier, we'd noticed that the flesh of the wound was quite red.

Dan and I were now due to look at my leg again, so I went back to the tent. With my rain pants off and the green zip-offs lying in a heap on the tent floor, I carefully unwound the bandage from around my thigh. It was darkened from use and stained where blood had soaked through. I peeled back the layers and removed the last bit of Ace bandage until I could see the ocular dressing over the main punctures and lacerations. Carefully, I removed the dressing. The main puncture hadn't improved, and the skin was flushed a dull red. Inside the wound, the raw flesh shone brightly, and deep inside was an accumulation of yellowish pus.

Seeing the obvious signs of infection, Dan checked thoroughly to see how bad it was. He told me I needed to do my best to get rid of the infected fluids inside the wound. We didn't want to waste the good gauze cleaning the pus, so I grabbed my thick blue PackTowl. Warily, I placed my hands around the holes in my thigh. The skin seemed to hum with anticipation, as if it knew what was about to happen. I tried to calm myself and focus. I tried to objectify my leg and objectify the task. *I am to press around the wound to remove pus, a necessary task*, I told myself. *It may hurt, but it has to happen.*

When I laid my hands on my thigh, the humming switched to a background buzz like cicadas. I felt the deep warmth of my skin. Tentatively, I pushed. There was no immediate pain, so I pushed more. Suddenly, it hit. I pulled my hands away and let the pain pass. It was like pushing on a really bad bruise. Once the sensation had subsided, I placed my hands again and, knowing what to expect, set to the task of draining my thigh. I pushed, and the pain came. My breathing quickened, and I gritted my teeth as thick, yellow pus came out, looking like curdled cream. Keeping pressure with one hand to avoid accidentally sucking the stuff back in, I grabbed the PackTowl and dabbed at it. It stuck easily to the towel, and in a moment I was ready to push more out.

I spent ten minutes squeezing the bear bite as if it were a giant zit. My hands shook as I worked, and I would have to stop and take a break before setting back to it. In the end, the deep blue PackTowl was dotted all over with thick smears. I'd gotten out all that I could. Deep irrigation washed out still more. Afterward, the pain from pushing lingered. I felt queasy. The sight of the PackTowl, the sight of the exudate oozing from the wound, and the ongoing hum of pain made me want to vomit. We carefully covered the wound. I tried to relax and calm the throb of pain in my leg, and the nausea in my throat.

Dan and I would check the wound in an hour. If it still looked bad or worse, he would call Camp.

The hour passed, my queasy stomach calmed, and the pain ebbed slightly. Dan and I went back into the tent and looked at the wound. It hadn't improved. We wrapped it up again and left the tent. I waddled around the campsite while Dan went to get the satellite phone.

Fifteen minutes later, Darin came up to me. He'd been talking with Dan, who had just hung up.

"You're going to get a present," Darin said.

"What?" I said.

"Dan'll bring it to you," he said, finishing with his characteristic bemused side smirk before stepping away.

Dan was headed my way, his hand clenched in a fist. When he met me, he held his fist out and turned it palm up before he opened it. In his hand were three horse-size pills—two white and one a bright pink. I supposed two were antibiotics, and I had no clue what the third was. With his free hand, Dan pointed at the two white pills. "Vicodin." He pointed at the pink pill. "And Erythromycin."

"But what are the Vicodin for?" I said.

He sighed, lowering his hand slightly. "They say I have to open up the clotted wound and irrigate it again, a lot, tonight. That's what the Vicodin's for."

I took the pills and tried to swallow my nervousness along with them. As they dissolved in my stomach, we ate dinner. Despite my palpable apprehension over the coming irrigation, I was still hungry for the hot, cozy meal.

With my belly full and my mind just a touch off-kilter from the narcotic, I hobbled around the campsite. When I was a little way outside the camp, a great herd of caribou suddenly dotted the horizon. They moved silently, silhouettes against the gray. The whole lot moved at the same slow grazing pace, as if they all were on a conveyor belt. They grew larger and larger as they came toward camp, walking within fifty yards of me in clumps of about twenty head. There had been many groups over the course of the day, but this was the most I had seen at one time. I watched them with a hesitant fascination. I knew they were caribou and not to be feared, yet I worried they would turn on me. I decided I'd try to scare them off if they came much closer. At that moment, the first group turned sharply away toward the next point of land to our east. Whole groups of them came toward me before veering off.

Our campsite marked the closest piece of land to the island just offshore in Thirty Mile Lake. We must have camped on part of their migration route. Hundreds of caribou ambled up, then turned from our campsite only to go a quarter mile down shore to the next point of land. From there, they slowed at the water's edge, bunching up like penguins at the edge of the ice. Then the first caribou stepped into the water, and again, moving as if on a conveyor belt, the whole group waded in. At this distance and with the buffeting of the wind, I couldn't hear them, but I saw the water splashing as they broke the surface. Once they were in, they sank up to their necks, so that just heads and huge antlers poked out, like a whole row of trophy mounts scooting across the lake. They moved together at the same speed across the water as well, rising up from it without slowing as they walked onto the island. Once up, they paused and shook the cold water off, halos of droplets forming around them like clouds. I watched more and more of them as they passed me, crossed the river, and disappeared into the distance on the island.

This migration happens each year, and here we were, quite by chance right in the middle of it. I finally turned back into the merciless wind, toward the tents, not yet ready for the next irrigation but not wanting to delay it any longer.

Dan was ready, and we retreated to the empty tent. The rest of the guys were in the other tent, and Dan warned them about what we were about to do.

I'd been through irrigation before and knew that I'd probably yell. I felt bad for Dan and the other guys. I couldn't imagine listening to me, especially without seeing what was happening.

Soon, I was laid out in the tent, my dressings removed and the wound open. Dan had already prepared a one-liter Nalgene of irrigation solution as the Camp health officer had prescribed over the phone. Dan was almost as uncomfortable as I was, his six-and-a-half-foot frame scrunched in the tiny tent, and a gruesome task ahead. His headlamp shone on the bite, and he had the syringe in hand, poised to start. In all this, it was his orange-and-yellow Hawaiian shirt that seemed most out of place to me.

I tried to steady my breathing. If I tried too hard, I would start to shake. I needed something to focus on while Dan worked, somewhere to channel my energy, to divert my flinching to, somewhere to direct my pain. I concentrated on the gear loft in the tent and saw my Leatherman hanging in the mesh. I looked at the blue carabiner attached to it, and the dark-red sheath. The leather was polished from use and hardened from repeated wetting and drying in the cold rivers and lakes of Canada. I realized the sheath might be perfect. I pushed up just enough to grab it, then dropped back down. I looked at the burnished leather. It had gathered a few scratches here and there, but it wore them like badges of honor. Removing the carabiner but leaving the Leatherman inside, I put the sheath between my jaws and clamped down lightly. It absorbed the bite well, giving enough so that my teeth wouldn't crack, while the rigid steel inside kept my jaws from cramping. I released it and looked at it. Tiny tooth marks showed on the corner.

When Dan asked whether I was ready, I said I was, and clamped down on the sheath. He lowered the tip of the syringe to the small opening in my thigh. The tissue looked as angry as when we squeezed all the pus out earlier. It was a deep, gory hole. Under pressure, it oozed blood, but it wasn't actively bleeding. Dan pressed on the plunger. Water spurted, splashed at the surface, then found its way into the puncture. For a long moment, the wound filled with water and none came out. It wandered inside my leg, finding every open crevice as it made its way deeper than Dan or I thought it could.

At first, it didn't hurt. The pain was distant, like touching a thick scab. Then it grew. I'd been expecting it, but the way that it grew surprised me. Pain had waited just long enough to make me think that it wouldn't come. Then, as soon as I thought I might be in the clear, it hit.

The cold water had reached my nerves. I clamped down hard on the Leatherman.

After a moment of irrigation, we saw the action of the water again, though not where we expected. We'd been expecting to see the water fill the wound and begin pouring out whence it came. What we saw surprised us both. Next to the puncture were the surface wounds where the bear's tooth had lacerated my skin, and underneath these, forming a bruise an inch and a half square, was the terrible compression wound I'd received from the same tooth. It hadn't broken the skin other than the surface cuts, but the whole square had scabbed into a dark black plate. Now, the entire scab rose, bulging like a wart on a toad. The water had found the depth and the ends of my wound and spilled back out through the puncture, dripping down both sides of my leg. Dan's plunger reached its stop. The solution slowly oozed out of the wound, as the huge square scab seemed to give a sigh. Meanwhile, Dan filled the syringe again. When the wound had wept all the solution it was going to, he aimed the syringe again and emptied it. Then he filled it again. And so it went: fill, spray, bite leather; fill, spray, bite leather.

By the time he was done, we both were exhausted. We sat still and tried to catch our breath. After a moment, he lowered the syringe and I took the Leatherman out of my mouth. The sheath was arced with tooth marks, though none had punctured the leather. I set it aside and stared again at the ceiling. It felt as if the Vicodin had done nothing. But I imagined how awful it might have been without it.

We'd been instructed to mark the progress of the infection, and Dan traced the red halo around my wound with a Sharpie. It looked like a cartoon splatter of paint—a large blob with bits sticking out here and there. We cleaned the wound, dried it, and applied the dressings.

I got my clothes on, and we cleaned up the tent, converting it from operating theater to sleeping quarters. We tried to dry my stuff that had gotten wet, and clambered out of the tent.

Opening the door was like peering out of a cave. We suddenly felt and breathed a freshness we hadn't even realized was waiting for us.

The sun had broken through a thin rift in the clouds, and it spilled onto the ground like molten gold. The grasses shone a bright yellow in the radiant sunlight as they shook with the wind. Behind the bright light, the dark of the gray-shrouded tundra stood out in stark contrast. The ray of light seemed to be hitting only our camp and the bit of tundra around it.

There was no good way to dry my sleeping pad, so Dan flapped it in the stiff wind. All the dimples, each a half-inch deep, had trapped the liquid. The wind was still vicious, and with Dan holding one end, my pad flapped wildly in the sun. I sat in the tent with the door open and imagined the pad slipping out of his hand and tumbling into the water.

Seconds later, the sound of flapping stopped abruptly. "Oh, shit!" I heard Dan say. Then came the soft splash. Dan leaped over guylines to get to shore and snatched the drifting pad out of the lake, saving me from sleeping on the ground for the rest of the trip.

After Dan had retrieved the pad, dried it once more, and returned it to the tent, I made my way through the door and into the light. With my boots loosely tied, I hobbled to the shore, near where the pad had splashed down. I wandered from there to one of the glacial domes of rock on our point. Staring at it, I decided I could clamber up the few feet, found a route, and carefully climbed up. From that small plateau, I could better see the wide expanse of Thirty Mile Lake and its islands. Some caribou were making their way to the crossing, and some had already crossed and were far away on the island. I could see the rolling dark clouds in the distance and felt the warm glow of the golden light. It was nearly flat, and one of the most simply beautiful vistas I had seen.

I was also looking at ground we needed to cover, a distance we must travel to get to Baker Lake and home. We'd been stuck at this site for two full days and nights and were coming up on our third. Our bid to catch up with the Femmes, resupply our med kit and satellite phone batteries, and regroup was seeming less likely every moment that we sat and did nothing. Meanwhile, my leg festered.

We'd been caught in the trap of Thirty Mile Lake. Like everyone, we'd paddled here only to get shut down by wind and rain. It would let us go eventually, but what would my leg be like by then? What would *I* be like? There was no possibility for evacuation during this storm. Only after we got stuck had my leg started to show signs of infection. Now I hoped for the antibiotic to work. It should; that's why we brought it.

I knew there were graves on the lake, and now, looking out across this great expanse, I felt their presence. I imagined the spirits of Thirty Mile Lake holding on to travelers as they passed through, holding *us* to account for disturbing their rest. I felt a pull in my stomach as emotion welled within me. It made its way up my spine and throat, sending chills over my neck and scalp and forcing my mouth open as words spilled out. They were whispers. I could barely hear them myself, much less speak them. From that rock, I pleaded with the spirits. I asked them to release us. I could feel my leg, too warm and disturbingly numb. The wound wasn't improving. We needed to move.

"Please," I said, "let us through …" My words disappeared on the wind, whipped out over the open expanse and across the length of the lake.

I watched for a long while until the glow had passed and gray returned.

> There have been joys too great to be described in words, and there have been griefs upon which I have not dared to dwell; and with these in mind I say: Climb if you will, but remember that courage and strength are naught without prudence, and that a momentary negligence may destroy the happiness of a lifetime. Do nothing in haste, look well to each step, and from the beginning think what may be the end.
>
> —**Edward Whymper,** *Scrambles amongst the Alps*

CHAPTER TWENTY-ONE

Surgery: Day 34

Each day we spent at the campsite, the caribou migration grew. Our third morning was no different. The herds became bigger and passed through more frequently. At one point, I looked out across the tundra to see caribou all along the north horizon, starting in the northwest, descending into the lake to the east, and disappearing in the distance of the island in the southeast. Predictably, each group got close enough to see that our site was occupied, then turned east toward the next point, where they would cross. I found watching them in the water fascinating. At times, hundreds of little bobbing heads were drifting across the lake, only to emerge as whole animals on the other side. I felt lucky to have ended up at such a crossroads.

The winds had calmed slightly, and Dan decided it was good enough for a sortie to see if we could find any more of the lost equipment—one last search for the canoe. Everyone but Auggie and me would go looking for it. The two of us would stay at camp. Auggie was exempt given his previous service, and I was exempt because of my leg. I was hopeful the antibiotics would begin to kick the infection soon, though. Auggie and I were to strike the tents and prepare the canoe packs in hopes that the weather would hold or improve and we would be able to leave when the others returned.

Dan, Darin, Mike, and Jean all climbed into one of the two remaining canoes and pushed off from shore. Auggie and I had lingered in the tent, and now we stepped out to see them off. We just caught sight of the four of them in the seventeen-foot boat as it rose and fell over great waves. They were on their knees, paddling in unison into the distance. They looked like some odd, colorful creature paddling with its four flippers over the whitecaps. They went amazingly fast. Four strong motors and no dead weight.

Before we knew it, the canoe was out of sight and the two of us were left in that vast space. *So long as there is another person*, my subconscious said, *it can't happen again.*

It was hard to remain idle on an expedition like this, so we set to cleaning camp. It was almost a compulsion to be doing something constantly, for fear that if we didn't, something bad would happen or we'd be unprepared. Also, there was no shortage of work. The oncoming night, departure, storm, wild animals—whatever it was, if we didn't prepare for it, we would inevitably be worse off. Surviving with basic means takes a lot of effort, and if we didn't do the work now, we would have to do it later, probably at a less convenient time.

We picked up all the bits of gear that seemed to have exploded from our packs. We removed the poles and flattened the tents, carefully folding each into a strip, rolling it into a neat bundle, and stuffing it into its bag.

At eight forty-five, we started breakfast, and almost immediately, the four-man canoe appeared on the horizon. They closed the distance quickly and reported that they'd found no canoe, no blaze-orange daypack, nothing. They had tried, and that was all they could do. Those things were truly lost.

By the time they were out of the boat, breakfast was ready. I took my second antibiotic pill. I would take the third with lunch and the fourth with dinner. The thirtieth pill would be at lunch on my way home from Camp in Minnesota. It was a mind-boggling thought. I could hardly imagine home, it was so far away in both time and distance. It felt like a purely abstract concept, and we had a long way to go before we got to that last leg of the journey.

Having finished eating, we stashed all the cooking gear in the wanigan and began packing our two boats. We had to place each piece of gear carefully. Not only must we fit all our equipment into the remaining two boats, we also had to squeeze another person into each. We got creative, standing packs up, stowing smaller barrels under seats and in bow and stern compartments, and filling the boats to the gunwales. By the time we were done, they looked like warships, with great sharp angles and big blue smokestacks.

I made my way to the duffer seat of one of the boats. With help, I lowered into the tiny compartment. It was seriously cramped. My legs were bunched up and spread wide, as if I were trying to sit cross-legged in an inner tube. It was uncomfortable, but I would manage.

Once I was in the boat, Mike got in the front, and Dan climbed in right behind me in the stern. We pushed off from shore and felt the familiar bob of water as it bounced the boat.

Leaving was a new sensation for me and, I think, for the rest of the guys, too. I hadn't paddled at all, and they hadn't been in loaded boats in the past two days and three nights, but the biggest change of all was how overloaded we were. With all the weight, our canoes sat low in the water, with only a few short inches between the gunwales and the lake. Compared to the six or more inches we were used to, that was almost no freeboard. By this time, we all were skilled paddlers, but it left little margin for error and made us appropriately nervous.

As we passed the caribou crossing at the next point, a herd of more than two hundred was gathering onshore, watching us as they waited to cross. Paddling silently, we looked at them, and hundreds of eyes looked back at the strange aquatic creatures floating past.

We watched each other quietly, the caribou's steamy breath hanging in a dull cloud over the dense forest of felted antlers. When we pulled up on the shore of the big island, I looked back and saw they had started their crossing.

We had landed at the island marked with graves and great inukshuks on the Femmes' photocopied map. From where we beached, we could already see one inukshuk. The tall stack of huge stones stood like a proud giant, commanding the grassy landscape around it. We hauled up the boats and walked toward it. As with everything in this vastness, it was farther away than it looked, and we marched for quite a while.

When I finally reached the site, I was a good way behind the others, who were already on their way to the next. I relished the sudden solitude. Alone with the inukshuk, I could see it more purely than if I were with the group. The bulbous blocks of stone loomed high over me. At least ten feet tall, this was by far the biggest inukshuk I had ever seen.

Many of the inukshuks in northern Canada are built of flat, wide stones, to look like people or other creatures. This one wasn't like that at all—more like a monolith, huge and solid. Its base was an enormous block of pink granite that would have taken several strong bodies to move. The stones got progressively smaller all the way to the top, where the smallest were still huge and looked immensely heavy. Each stone was trapezoidal, with its widest face on the outside, as if each were a keystone holding something in. It looked as if the structure might collapse outward if pushed from inside, but would withstand any onslaught of wind, snow, and ice. It was a powerful structure.

I photographed it and marveled as the bright yellow, red, dark blue, and black dots that were my companions shrank into the distance. I finished my photography and paid my respects before turning to follow them.

Ahead of me on the tundra, my traveling companions had stopped. They stood in a semicircle, almost motionless, around a low pile of stones. Drawing closer, I could see a rock enclosure about three feet tall and wide, and five feet long. In between the stone walls was open air, all the way down to the soggy ground. My companions were standing around what looked to be a grave.

I looked in past the walls of stone. Lichen clung to everything, covering the stones in a shag of orange, green, and sage. The ground was so mottled with colors bright and dark that at first glance, I nearly missed the white glint of bone. My eyes adjusted, and I saw the clear domed outline of a human cranium. I followed its curves down to the empty eye sockets, the hollow nose, and the upper jaw. It was an open grave. This was old. The bones, like the stone wall around them, were covered in lichen and moss, showing stark white between the colorful living growths.

The other guys left the graves, but I lingered. Just yesterday, I had stood weakly at our site, pleading to the spirits of Thirty Mile Lake to let us pass, to give us a break in the weather. I felt I was looking into the face of one of the entities I'd been appealing to. It was with great respect and thanks that I looked into this grave and the graves around it. I felt a strong connection with this sacred site. The skeleton was almost an abstracted form, its connection with the human body hard to imagine. It struck me that before this grave, there had been a difficult life on the tundra, hunting caribou and living in this vast, stark landscape. I imagined the life that had preceded this lichen-covered heap. Just as we live our lives, so had they. I had nearly become bloodied bones myself only a few days ago. I felt lucky, grateful.

I stood there for a long while before I could form words. "Thank you," I said. I opened my mouth again as if to say something else, but there was nothing more to say.

I turned back to follow the group toward our two waiting canoes. It was a windy day but manageable with no rain. When I got back to the canoes, I asked if I could paddle instead of sit in the duffer's seat. Scrunching up in the tight space at the bottom of the boat had been painful for me, and the Ace bandage stretched and rolled into a thin sharp band that pressed on my wound. No one had a problem with this, and I switched with Mike. Everyone else loaded into the boats, and we pushed out into the channels of Thirty Mile Lake.

Paddling was much more comfortable than duffing. Sitting upright and pulling the blade through the water took almost no effort from the lower half of my body. I felt free, I didn't feel pain, and I was contributing to the group as much as anyone could in a canoe.

Dan, as always since the attack, was in the boat with me. He was trying to keep me safe and also monitoring me for any change in mental or physical status. I couldn't imagine the emotional and physical strain he was going through. He did well, never showing that stress. His patience decreased, and his tolerance for bullshit ended, but given the situation, it was understandable. When we were paddling, we all felt some relief. All of us knew that to keep going meant getting closer to the Femmes, closer to Baker Lake, and closer to medical care for me. We were doing what we needed to do. Being out on the waters of Thirty Mile now felt wonderfully freeing, like stepping out of the stale air of the tent into fresh sunlight.

We paddled a long way down the lake, starting near its western beginnings and aiming all the way to the far eastern end, some thirty miles away. At seven thirty in the evening, our prearranged time, we stopped to check my leg. We pulled off to the north side of the lake and beached the canoes on a small peninsula near a large bay. The boats scraped the shore, and we climbed out.

Dan grabbed the med kit, and the two of us walked away from the canoes and the rest of the group. There was a sharp rise not far from where we'd landed, and we set up behind it to provide me with some much-appreciated privacy. We unwound the Ace bandage. The day before, we had seen the dark black scab of the terrible compression wound next to the puncture rising and falling with irrigation fluid, and we had pushed out the foul pus. With luck, by now the antibiotics were beginning to fight the infection and there would be improvement, or at least no noticeable change for the worse.

The gauze stuck to the wet edges of the wound, and Dan pulled delicately until it was gone. What we saw underneath didn't look good. There had been one puncture wound. Now there were two large holes, with a big, sticky white mass between them. We weren't expecting this. Knowing we needed to get to the end of the lake before we stopped to do anything, we quickly wrapped it again. Soon, the gauze was secured with the bandage, and I was clothed and ready to hobble back to the group.

Dan and I went back to the boats, and he grabbed the satellite phone.

He stepped away and, when he was out of earshot, dialed Camp. He stood with his back to the wind and updated them on the grim turn my leg had taken.

After a few minutes, Dan lowered the phone and powered it down. He had new instructions and would be calling Camp again first thing in the morning. The health officer had advised him on what to do next. He would be exploring, debriding, and aggressively irrigating my wound as soon as we made camp.

Dan rifled through the med kit and presented me with two more huge Vicodin pills in his palm. I looked at them and then him.

"We haven't had dinner yet," I said.

"I know," Dan said. "We need to work on this as soon as we get to camp, and we need these to kick in by then."

Reluctantly I took both, followed by a big swig from my water bottle. I'd never taken any hydrocodone on an empty stomach, much less two at once. We were going to get on the water, paddle like hell across a huge bay, and make camp on the flats on the far side, the eastern end of Thirty Mile Lake. This could be interesting.

I carefully got in the boat, and the rest of the guys pushed us off and hopped in. We turned and followed shore for a while, paddling hard into the wind. All of us were strong paddlers, and when we put some spirit in our stroke, we could fly. Today was no different, wind be damned. We rose and fell over crests and whitecaps, moving steadily eastward.

It wasn't long before I felt the effects of the Vicodin. Earlier, after taking it for the irrigation, I had felt the drug only distantly. It was always a foreign sensation—a little disorientation, a slight loss of inhibition, especially with words. I had never really felt a wave of comfort or laziness, or a high from the drug. What I started feeling in the boat was none of those things, either; I started to feel woozy. It was a drowsy dizziness, almost a clumsiness. I didn't feel as if I were about to fall out of the boat or get so disoriented that I would try to stand up. I did feel less precise, as though I might start bumping my hands against the gunwales if I didn't watch it. Dan and Mike asked again if I wanted to duff and got a resolute no. Duffing was uncomfortable, and I wanted to focus.

I never get carsick when I'm driving, but on mountain roads, if I'm not behind the wheel, there's a distinct possibility that I'll get queasy, especially if I've got my head down in a map or a book. In the bow, I could provide most of the "motor" for our canoe, and with the Vicodin, I found out that I could push as hard as I wanted. I was up with adrenaline in anticipation of the invasive procedure I was racing toward. My muscles weren't tiring as I had expected. The harder I paddled, the more my shoulders numbed, and I wasn't tiring. It was as if lactic acid never even built up. My breathing quickened, and I pushed our boat forward, propelling us into an unknown, uncomfortable future. The sooner we got there, the sooner it would be over, and the sooner the nervous anticipation could leave my mind.

Near the end of the lake, we approached a large peninsula to the north. Behind it, a huge bay angled off to the northwest. I scanned the water just beyond. We were still a good mile away from the crossing at the open bay, so both the peninsula and the water beyond it were mere slivers on the horizon. I squinted at them. The point was brown and green, with dark splotches of stone along the water. The water was dark, steely blue and gray, but just past the point, it seemed to change to a lighter shade. Also, it appeared to switch from the low, flat waves we were paddling through to something taller. I couldn't make it out at first. It was blurry in the distance, and we were constantly moving. It could be a mirage. But the more I looked and the closer we got, the surer I was that the water changed there. A few more strokes, and I could see more clearly. Past the peninsula, the water suddenly whipped into tall, white-topped waves.

"Dan, we've got some big waves up ahead," I said.

"Where?" he asked, leaning his head to the side to see around me.

"Up there," I said, pointing with my paddle toward the peninsula. "Just at the point. Past that, the water seems to whip up into big waves."

"It looks like ice," Dan said, which would be worse than waves.

"I think it's waves."

By this time, the Vicodin had really begun to kick in. He had noticed and was likely taking most of what I said with a grain of salt.

The closer we got, the clearer the waves got. Not ice and not a mirage, they were rising and falling individually, sporadically, without the sense of order I expected. The oddest thing about the waves was that they started along a crisp line that ran south across the lake. We were about to cross a force field into some sort of maelstrom beyond.

We closed in on the waves and scouted them as we would a rapid. They stretched as far as I could see to either side, a solid wall of unknown thickness. We had to push through if we could, though, so we aimed straight in, keeping our crossing to the far shore as short as possible.

"It's just like white water!" Dan said, raising his voice for me and the other boat to hear over the thundering waves. Everyone else dropped to their knees, while I sat awkwardly upright in the bow.

Have to cross this, I thought to myself. *Have to paddle. Can paddle.* I dug my paddle in, and we crossed the force field into the cacophony of waves.

The canoe bounced up and down without order or rhythm. The wind whipped in from across the deep bay to our left, tossing the water into mayhem. We'd been riding the steady, slow current of Thirty Mile Lake all day, and suddenly it hit this crosscurrent. The wind had been funneled by the body of the lake, too, parallel to the current, so that we were in the middle of conflicting crosswinds, as well. Waves bounced randomly, and the farther we pushed, the taller they got.

Underneath the boat, I could feel up-currents like those on the river. I kept paddling, kept pulling on. The waves rose until some were five feet from crest to trough. We bobbed and fell so steeply that the paddlers of one canoe saw the other only in glimpses of bow or stern cresting a wave. Most of the time, the boat tilted precariously up or down, leaning to one side or the other. Dan worked hard to keep us straight and level while I sat up at the front like a tall, heavy mast. I just kept paddling.

If I could have felt them, my shoulders and back would be burning, aching for a respite. I wanted to get off this chop and into camp. I could sense that things were getting worse with my leg, and I worried about what would come next. I didn't want to get stuck somewhere again. We just had to cross this violent field of waves.

We kept pulling, the only breaks occurring for the instant when we switched sides. As we neared the far shore, the waves calmed. We paddled hard, the momentum of the crossing spilling out so that we beached high up on the far shore.

We all were out of breath, but the light in the sky was dimming, so we quickly set to making camp. I concentrated on helping with one of the tents. I knew we were building an operating room.

The shore had small bushes and low plants that shook in the wind. I stepped over and around them as best I could, but between the little shrubs, the Vicodin, and bending over to place stakes and poles, I grew dizzy, and my steps faltered. By the time we were done, I needed to sit and let the spinning in my head slow. The whole time on the water, I hadn't felt the least bit seasick. Now, on dry land, I could barely stand, and I was uneasy about what would come next.

The tents were up, staked, guyed, and ready. One was full of sleeping gear; the other had my lonely sleeping pad in it. Before we entered the low-ceilinged dome of the Prophet, Dan spoke to the rest of the guys. They paused what they were doing as he raised his voice over the wind.

"Don't come by the tent until we come back out," he said. I could hear uneasiness in his voice, though he masked it well.

They all nodded and said okay.

I hobbled in and worked my boots off, and Dan came in right after me. The modest light from outside was much dimmer inside. With no gear and only two bodies in the tent, it felt spacious, though the ceiling still felt low, especially for Dan, who crouched uncomfortably. He and I were quiet, silently getting ready. Soon I had the stained elastic wrap exposed. Dan mixed a potion of irrigation solution in my designated Nalgene bottle.

I awkwardly sat myself up and worked at the bandage. Unwinding it was like peeling back a curtain. It hid what was underneath, and getting to that was painfully slow. In a heap on the floor of the tent, the bandage looked more like a shed snakeskin than a medical supply. The gauze didn't look much better. The wound was oozing, and the white packing had been thoroughly stained.

Dan had finished mixing his solution and had gloves on, and more gauze and the irrigation syringe ready beside him. He turned to me and flicked on his tiny headlamp. The LEDs were like a row of stars, like Orion's belt. The batteries, weak from over a month of intermittent use, were not as bright as we wanted, but they would work.

The glow landed on my wound. Around the gauze, the pale red of the infection was unchanged. With gloved hand, Dan carefully pulled the gauze from my leg. It was a distant sensation, and I felt the cool air more than I felt the gauze being taken away. I watched as he peeled it back.

When we saw the wound, Dan huffed out his nose, and I sighed a resigned "*Shit.*"

An hour before, we had seen the puncture, and the new opening across the sticky white mass. In the hour since, a third hole had opened on the other side of the mass. This was not good. There were now three openings, a triangular constellation around the wound. Either the infection was moving fast, shrugging off the antibiotic, or the tissue was so damaged and fragile that it was falling apart as I moved. Either way, it was bad news, and we needed to clean each hole thoroughly.

Dan filled the syringe, and I lay back on my sleeping pad. He put the narrow tip near the opening of the original puncture, his thumb steady on the plunger. By now well practiced, he smoothly pressed it down. Water shot in a jet straight down into the red and yellow flesh.

He hadn't put more than 30 cc in when he stopped. "Whoa," he said.

"What is it?" I said, breaking my unsuccessful meditation and rising up on my elbows.

"When I spray water into one hole ...," he said, looking first at my wound and then to me, "... when it sprays in the one, it comes out of all three."

"What?" I said, at first confused, and then understanding as he explained it to me.

"They must be connected," he said. "There must be an opening inside."

"Fuck," I said.

This didn't change the fact that we had to clean it; it only reinforced the urgency. I watched now as he irrigated. He sprayed into the main puncture, the one where the bear's canine had pierced my skin and pushed through to the gums, ripping past my femoral artery by a quarter inch. Entering there, the irrigation solution came bubbling out the other two holes, each an inch and a half from the main puncture. Something had changed within the wound. When the bear's jaw clamped down next to the main puncture, it had compressed my tissues so hard that it collapsed all the blood vessels. No capillaries, veins, or arteries in that patch of my leg would ever again receive blood flow. We were looking at the direct result of a compression wound. Just over five full days had passed since I came running down that ridge. We had seen the bruise turn to scab, and the thick scab become a soft, blanched mass. We hadn't realized it, but we'd been watching tissue death. The wound was now necrotic, dead and rotting, and underneath the surface it had formed an abscess with sinus tracts that now reached the surface.

Dan carefully and thoroughly irrigated the main puncture wound before moving on to the other holes. Into each opening, he sprayed irrigation solution, and each time, the other two holes wept. As he worked, the flow from one hole to the other two increased, switching from a subtle bubbling ooze to a steady flow nearly at pace with the water going in. The detritus was being flushed, the opening was growing, and we could do no more than keep cleaning it out and see how far it went. The wound had opened up and was now a roofed cavity—a dammed infection.

Dan finished irrigating and sat silent for a long while.

He turned to me. "Alex," he said, pausing a beat before continuing, "I'm going to have to cut that out." Each word was clear and firm. He didn't stumble, stutter, or otherwise betray any hesitation except in his tone, which was apologetic. But he knew that it must be done.

My mind churned through the words, connecting the meaning and tying it to anticipated action. *Cut that (part of my leg) out (of my leg).* Slice, snip, stab—surgery. I shuddered inside. I stared flatly at him for a moment. "Really?" I said, concerned but genuinely curious. "I mean, if you have to." I didn't want to unwittingly dissuade him from doing something important. I knew he didn't want to cut into me. I wasn't thrilled at the prospect, either. In fact, I was nervous. Irrigation is one thing, but unanesthetized surgical wound debridement is something else entirely.

I looked up to the gear loft of the tent and saw my Leatherman hanging heavily in the no-see-um mesh. I reached up and pulled it down, for a moment thinking of offering it for use as a surgical instrument. I quickly thought better of it. Given the proximity of the artery pulsing just next to the open wound, I didn't want something too sharp, long, and lethal, knowing I was likely to flinch. I turned the heavy tool and leather sheath in my hand. The few tooth marks it had gotten during irrigation the day before were just visible.

Dan rifled through the med kit and inventoried the available tools. He looked at the Leatherman in my hand, then went through the boxes and bags in the huge first-aid kit. There were no scalpels. This kit had not been assembled for this kind of procedure—at least, not one of this scope. Dan was a wilderness first responder (WFR), schooled in remote care and transport of patients. Half of WFR training is in how to improvise medical supplies and techniques with what you have. A tool's intended use is not its only use. As he looked through the kit, he paused. In with the bandages was a small, shiny hook of steel: the shears for cutting medical tape. The blades were small, smooth, somewhat sharp, and the handle had two small, easily controlled finger holes. Most importantly, the blades were protected by a wide, dull shovel tip. Even if I flinched, the scissors wouldn't cut anything except what Dan put between the blades. He pulled the shears from the bag and began cleaning them thoroughly.

I looked at the Leatherman sheath and turned it so my new bite marks would align with the ones already there. I set the heavy leather-encased multitool between my lateral incisors and bit down hard enough to hold it there.

Dan finished cleaning the scissors and turned to me now. He carefully explained what would happen next—he planned to unroof the abscess and debride the wound so it wouldn't fester. His face was deeply shadowed behind the dim glow of the headlamp, and I watched the subtle shine of his eyes as he talked. When he had finished, he turned to my leg, and the harsh light illuminated the awful sight of my thigh. It shined grotesque, terrible colors in the light. I laid my head back and looked at the dome of the tent. Dan held the clean steel in his gloved hands, ready for his first surgery.

"Dinner's ready!" Darin announced from ten feet outside the tent.

Neither of us had heard him approach, and we both let out the breath we'd been holding in anticipation of the first cut. Dan snapped at him to go away.

I lay back again and focused on stilling my body. I was worried that the pain of the scissors shearing my skin would make me flinch and, shovel tip be damned, the tool would dig into my thigh. I breathed deeply, anticipating pain. I felt nothing. I couldn't imagine this not hurting. Still staring up at the tent roof and not wanting to move, I spoke through the leather between my jaws. "Have you started yet?" I asked.

"No," Dan said, a little annoyed. "I'm getting ready."

This was an altogether new experience for him, and he was the one at the controls. He was the one who really had to focus.

"This isn't easy for me, either," he said.

"I know."

I had to keep still, while he had to work precisely with tools on a preferably still subject. I'm sure he had never cut into a person before. I lay back again and stared at the red fabric of the tent.

Soon, he put the steel to my leg and began cutting away the necrotic tissue. I breathed as slowly and smoothly as I could. I could feel it. It started as a dull sensation. I did my best to focus but could not control the shaking. All my attention went to holding my hips and thigh still, but my feet, hands, and shoulders betrayed me with their shivering.

He paused the blades. "Is it hurting that bad?" he asked.

"No," I said, my body and voice shaking, "I'm just really nervous"—nervous that it was going to hurt a lot.

This needs to happen, I told myself. *There's no point in thinking about it not happening, because it will; it has to.* I stilled, and Dan went back to his work. I could tell what was happening with my wound in an abstract sense. I knew that Dan was cutting into my leg. I could feel his maneuvering, but I could not feel the sharp sting I expected. That part of my leg was numb, like a sleeping limb. And as with a sleeping limb, I could feel sensation, as if from a distance. The pain was stuck there, trying to push through but not building up enough charge to get to my brain, and my leg tingled with the effort. I felt the dull scrape of the scissors like the smooth edge of a fingernail running along the back of my hand. I steadied my breathing, slowing my exhalations. I felt the scissors creep through, parting the dead flesh. Suddenly, a spark of pain as the blades went through one of the few remaining live nerves within the dead tissue. I gasped, and the leather groaned under the pressure of my teeth.

As Dan worked along, cutting off pieces of my leg, he picked carefully at them with the tweezers and deposited them in our small red biohazard bag.

Time moved slowly, and after what felt like a long time, I asked whether he was done. "I'm nearly halfway," he said.

"What!" I asked. *We still have just as long to go? I thought we were nearly done.*

"It's really dull," Dan said, his voice almost breaking "And ... the blades keep sliding and won't cut through."

I tried to push the added visual out of my head, but there it was, shiny steel on slippery dead tissue, trying to cut through it like trying to cut through the tough parts of raw chicken. *Breathing.* I focused on breathing. *In, then slowly out.* Spark! Another nerve, another sharp intake of breath. My fists clenched tighter with the pain. I was holding my breath now. *Let it out slowly ...*

Time seemed to slow so much that it might stop altogether. Then, miraculously, Dan was finished.

He carefully irrigated the wound again, cleaning up the last bit of his work. I still lay flat on the mat, staring at the roof of the tent. It was done. I was exhausted. The liter of irrigation solution was gone, and the biohazard bag had bits of my leg in it. I rose up on my elbows again to take a look. The sticky mass was gone. In its place was a wide void at least an inch and a half in diameter. It looked clean. There were smooth marks around the parts of the edge where Dan had cut, and the fleshy rippled edges and floor where the tissue had died or been eaten away by infection. I was struck with the similarity between the flesh inside my open wound and the look of a thickly marbled rib-eye steak. Deep red, with buttery-yellow veins and inclusions. It made me want to vomit.

The deep, wide wound had an indentation where it went deeper at the bottom corner. One of the hip flexor muscles, which had been partially severed by the bite, was in worse shape now, a quarter of its circumference taken by the initial wound and resulting infection. At the lower end of the wound, the abscess had begun to descend down the muscle. There, the mottled bottom dipped in a shallow cone. On the skin's surface just beyond, I could feel the muscle beneath. It was different from the surrounding tissue—hard, like the stiffness that surrounded the rest of the wound, but going much farther, creeping several inches down my leg. Luckily, there was no red streaking, no sign of the infection spreading beyond the lines we'd drawn the day before, but the red aura around the wound remained.

Before removing the roof of the abscess, I felt a sense of tightness, as if my leg were trying to breathe but couldn't. With the roof gone, I felt as if my leg could breathe again. I didn't feel good, and I was completely exhausted, but that suffocating feeling was gone. The bit of dead flesh was an evil thing, and I was rid of it. It was trash now.

We both were silent.

Dan cleaned up the tools and covered the wound with fresh gauze, dutifully wrapping the Ace bandage around the thigh.

I slowly got dressed and prepared to climb out of the tent. When Dan opened the door, the cold fresh air from outside flooded in and we suddenly realized how warm and humid it had become in the tent. Both of us were sweating and still out of breath, still feeling sick, still silent. The tundra had darkened into full twilight. From the rippled sky to the endless expanse of gently rolling land around us, and the smooth, cold lake to the west, everything was dull, flat shades of blue—even the tents and canoes.

Seeing us outside the tent, the guys came over tentatively, not wanting to approach if they were not supposed to. We made no motion to wave them off, and they came up to us. Darin gave Dan and me our cups, filled with dinner. It was now cold—some version of a gray, colorless mush with red cranberries that bled into the gray. I put a spoonful in my mouth and chewed it with disinterest. It rolled around and seemed to dry up more as I chewed. I forced two of these bites down and returned the cup.

"I'm sorry," I said quietly. "I'm not hungry." And I turned back to the tent. Dan did the same.

I methodically prepared for bed, changing clothes and climbing into my sleeping bag. I still felt the unease, the nausea hanging in the background. I tried to push it away. It was time to sleep.

This is a fine chance to let go, to "win my life by losing it," which means not recklessness but acceptance, not passivity but nonattachment.

—Peter Matthiessen, *The Snow Leopard*

CHAPTER TWENTY-TWO

Evac: Day 35

The next morning began like so many others. Dan woke us, and we all gradually regained consciousness. I opened my eyes to the darkness inside my eye mask and lay still, staring at the nothing before I pulled the mask off. The bright light startled my eyes, and I squinted hard against it. We all were awake but lying still and silent, thinking.

I thought about where I was, when it was. Then I remembered my leg. I thought of the hollow spot just below my hip. *Does it hurt?* I asked myself. *No, not really, no change.* I still didn't move. I felt like lead, and I liked the weight. I didn't want to break the spell, so I stayed there.

The other guys eventually got packed up. Soon, the other tent was cleared and taken down. We had packed up most of our stuff, but Dan and I remained in the last tent.

He shut the door and we set to work on my leg. I unwound the Ace bandage, around and around my thigh. Gradually, the gauze was revealed—a much wider swath of white to cover the new inch-and-a-half opening of my wound. Dan was ready with gloves and peeled back the bandage. The large opening of flesh was still red, still mottled, and still dipped down into my hip flexor at the bottom. It still looked like a steak, with extra marbling. Still looked infected. The tissue around it was still stiff. We weren't happy with the state of things, but we weren't any more worried than we had been last night. It was still bad news.

We packed it again with gauze and wrapped it with the elastic bandage, and Dan pulled the satellite phone from its protective hard case. When he powered it on, the screen lit up and soon displayed a lock with several satellites. The bars climbed until he had a good, strong signal. This, the spare battery, had two bars of energy remaining. It was almost empty.

Shit, Dan thought, *not much left.* He thumbed the number for Camp and clicked the green button to dial. Sitting in the tent, with his elbows wrapped around his knees, he watched the screen as the call began, then put the phone to his ear. The earpiece was so quiet that just a few feet away, I didn't even hear the ringing on the other end. I sat listening to the quiet bustle outside, the wind as it rose and fell, and the sound of our breathing.

"Paul, it's Dan," Dan said into the phone. He went on to inform Paul of what he'd done the night before, the surgery, how my leg had looked this morning.

I stared at the floor. I wasn't usually privy to these conversations and wasn't listening too intently at first. I was lost in my thoughts, not wanting to revisit the ordeal of the night before. *What now?* was the main thought in my head. I felt a sense of reaching the end of a road, a T, and I didn't know whether we'd go left, or right. I didn't even know whether there was a left or right, much less what lay in either direction.

Once Dan finished describing my leg, I began to listen more intently. I still couldn't hear the other end of the line, but I caught Dan's every word, trying to piece together the other half of the conversation. I didn't need to hear the other half. It was clear what was about to happen.

"Yeah," Dan said.

Silence.

"Okay," he said. "... right ..." He sounded tired, expecting what he was hearing, but not happy about it.

Dan paused, waiting for Paul to finish a sentence. "So, find a landing pad?" he said.

I felt a surge of emotion. There was a pull in my stomach, and chills ran over my body. The hair on my neck stood up, and I felt a tingle across my scalp. Goose bumps prickled around my torso and echoed out through my fingers and toes. I was getting medevacked. *That's it,* I thought. *We tried. Now it's over.*

Dan waited, still listening to the other end of the line. "Okay ...," he said slowly, as if he didn't quite agree with what he was hearing now. "No, I'm not worried about that ..."

Silence again as the phone spoke back to him. "No," he said, "I don't think that's necessary."

I was curious now. What was the disagreement about?

"Okay," he said finally, "I'll ask him."

With a few more words and after establishing the next protocol for call times, Dan said goodbye and ended the call. He lowered the phone, powered it off, and rotated the antenna back into the body of the Globalstar.

Both of us were silent. I stared blankly at my feet and the tent wall.

He put the phone back into its case and clicked it shut before turning to me. "I'm guessing you know what happened in that conversation," he said, as much a friend to a friend as a guide to a young camper.

I was looking down now, my view of him blocked by my hat brim and my hands as they held up my forehead. "So," I said, trying to keep my voice free of the emotion I felt pumping through my body, "I'm being evac'd."

"Yeah," Dan said. "The helicopter should be here within two hours."

Click. Just like that, my trip would be over. Thirty-six days of paddling and portaging and work, and suddenly, two hours from now, my ride would be here. We had paddled ourselves within range of Baker Lake, and there were aircraft available. Now a huge, thumping helicopter would rise off the tarmac and come get me.

I sat trying to process the sudden change, the inevitability we'd been pushing up against since the moment I crested that hill. *Damn it,* I thought, *we've tried so hard, and it still wasn't enough.* There wasn't any more we could have done. We'd been thrown an unimaginable challenge and been beaten down even more since then. In the past six days, I'd been mauled by a grizzly bear, we had evacuated camp because of roving musk oxen, we had set a night watch, we'd escaped the fog, I'd learned to walk again, we'd paddled miles of swift river and white water, we'd lost a boat and nearly all our food, against all odds we'd found our food again, we'd endured a gale-force storm for three nights and two days, we'd watched thousands of caribou in the midst of a great migration, we'd cleaned and treated my leg many times, we'd greeted the ancestors of the people whose land we were crossing, and we'd gotten to this site and done what we never signed up to do with my leg. We had done everything we could, risen to every challenge. And at this moment, a helicopter was being dispatched to come find me in the vast emptiness of the Nunavut tundra.

I didn't want to leave, yet I knew that I must. The condition of my leg had worsened and was beyond our means to mend. The antibiotics we were treating it with were not strong enough to fight this infection. This was the next step, and I would take it. It was what we needed to do.

I took a deep breath and let it out slowly. Finally, I turned to Dan. "So," I said, finding my words slowly and quietly, "what weren't you worried about?"

"Well, they were talking about posttraumatic stress syndrome," he said, "and they want someone to go with you."

"What!" I said, almost startled. *How can I be responsible for pulling someone off this trip? I'll be fine on my own; I'll be taken care of. It's not fair to make someone else leave. This is my struggle. I was the only one attacked, no one else, and no one's trip should have to be cut short because I need medical attention.* "I don't need that," I said, struggling to put my thoughts into words. "I'm … I'll be fine on my own. I don't need someone to come off the trip just to come with me."

"I know," Dan said, "that's what I told them." I could hear in his voice and see in his slouch that he was resigned to it already, resigned to the fact that I was about to board a helicopter with another of his campers, and the two of us would disappear from the trip. "Who would you want to go with you?" he asked.

Damn it, I have to choose, too? I have to pick whose trip ends today? I shook my head and sighed, trying to justify, trying to decide. Camp had the perspective we couldn't. They'd been in touch with experts, emergency personnel, health officials, lawyers, everyone you could think of—a full crisis team. We'd been in a tent and a canoe while this whole network of people and organizations had been scrambling. They knew what needed to happen. The equation had changed yesterday when my skin and muscle and nerves began to die and opened up from the inside. Camp knew that and had set plans into motion to have the helicopter ready come morning. We were suddenly fighting a different kind of battle with the infection in my leg, and the aim had switched from expeditious paddling to immediate evacuation. And they knew it wasn't something to send a seventeen-year-old to do alone.

I wanted to yell. But this was what needed to happen, and I must choose one of my friends to come with me so I wasn't alone. The faces of my companions scrolled slowly through my head.

"Well," I said—*I don't like doing this*—"probably Mike or Auggie." Their names came out in the last of my breath, and I felt deflated. I had just ended someone's trip, and I felt awful for it.

"Okay," Dan said, "I'll go talk with the guys." With that, he got up and climbed out of the tent.

Suddenly, I was alone, now truly in silence. His footsteps disappeared in the distance as he walked to the kitchen. By the time he got there, I couldn't hear his steps, much less his or anyone else's words. I was truly alone in the tent.

I could have sat there getting lost in my reflections, but I brought myself back to what needed to be done. My mind raced through everything, tried to process what had happened and what was about to happen. But thinking about it wouldn't change anything. I was about to leave.

What did I need to do before I went? I needed to pack up all my gear, everything that the group did not use.

I set to packing up my things. The silence in my head shifted abruptly to the anticipation of sound, as when a needle touches down on a vinyl record. The background white noise and steady whip of the wind buzzed as the stylus in my mind touched down on its own spinning LP. The silence at the start of a track began, and as I finished folding my sleeping pad and looked around the tent for my other things, I heard the clear timbre of a piano chord.

It was a song I had listened to when I was sad. And there on the tundra, it played as clearly as if my ears were hearing it.

As the song began, I continued looking around the tent. In every corner, there seemed to be something of mine—things I'd shared with others, things I'd kept to myself.

I took my Leatherman from the gear loft and gathered my sleeping gear. I pulled my water-damaged books from the reading dry bag. The raw paper edges of *The Snow Leopard* were stained in streaks like brown marble, and the covers expanded outward from repeated soakings and drying. Placing the tattered book into my bag brought a wave of sorrow. I was collecting my things from the group bags, as if erasing my presence and memory from the trip and the group. My emotions were irrational. I felt a sense of loss and failure, as if all our efforts after the bear attack had been in vain. I felt, too, a sense of foreboding, of a future unknown. The situation had been entirely out of my control during the bear attack, and it was horrifying. I had regained some sense of control as we left the island and moved east. Now, I was losing it again, and with this came an almost overwhelming sense of the unknown.

As I looked around, it started to hit me. I was about to leave. All this was about to be over. It made me immensely sad in a way that I had never felt before. With each item I saw and each thing I grabbed, the feeling welled up inside me.

Outside the tent, the song played on. Bits of my gear were everywhere: jackets stashed all around in different bags, my paddles, my fishing gear. I made a pile next to the tent.

Forty-two days' worth of food and gear in a modest pile, ready to be packed into the Twin Otter plane at Lynn Lake on July 03, 2005.

The group poses in front of the Twin Otter float plane before starting the trip. Left to right: Auggie, Jean, Mike, Alex, Dan, Darin.

The serene Canadian wilderness seen through the left propeller of the Twin Otter float plane as the crew flies north to Wholdaia Lake and the start of their trip.

Dan and Mike navigate a small whitewater set on the Dubawnt River.

The taiga forest of the Northwest Territories.

Darin and Mike take a break on the ice.

The Dubawnt River carves a deep, wide canyon as it flows around a huge S curve.

Navigating through Dubawnt Lake's ice and open leads.

The sun sets over Dubawnt Lake ice.

Mike takes the painter line of one of the boats as the group navigates around an island that was surrounded by ice overnight.

Dan, Darin, Mike, and Auggie navigate through ice leads on Dubawnt Lake.

The author tows a canoe across frozen Dubawnt Lake ice, holding his paddle to aid as a brace in case the ice fails.

A grizzly bear saunters into the distance, as seen across Dubawnt Canyon.

Dan and Jean walk across firm ice after their slog through candlestick ice on Dubawnt Lake.

A musk ox grazing across the tundra.

The group's first layover day on the flats in the middle of the Dubawnt Canyon portage.

The group stands on the esker marking their departure from the Dubawnt River. From left to right: Alex, Auggie, Dan, Darin, Jean, Mike.

Portaging the esker bushwack portage from an unnamed lake to Tebesjuak Lake.

The author proudly displaying his biggest catch from the trip.

Paddling across a glassy Princess Mary Lake on July 30th, 2005, day twenty-eight of the trip.

Auggie preparing freshly caught lake trout fillets for dinner.

Bedroom, rec room, changing room, and operating room—the MSR Prophet tents on the trip served as almost everything but the kitchen. Shown at the site on the west end of Thirty Mile Lake.

Dan looks east across Thirty Mile Lake during one of the first breaks in the storm.

Some of the first sunlight at the western Thirty Mile Lake camp.

Some of the thousands of caribou that migrated past the site while the group camped at Thirty Mile Lake.

Alex, Darin, Jean, Dan, Mike, and Auggie pose for their last group photo of the Hommes trip at the east end of Thirty Mile Lake on August 06, 2005.

Jean, Dan, Darin, and Auggie wave to Alex and Mike from the east end of Thirty Mile Lake as they depart by helicopter on August 06, 2005.

Dan, Alex, and Mike pose in the Menogyn parking lot after their Hommes du Nord trip. August 14, 2005.

As I went about the camp, I did not find the task calming. I grabbed my last stray things and hobbled to the pile. It was almost more than I could take. I was so sad to be leaving. Auggie or Mike had to leave, too, and I felt awful about that. The song built up, and I started to break down. I dropped to my knees at my pile, no longer able to hold myself up. I felt my jaw quiver and tried to hold back the wave of tears building deep inside me. I hadn't yet cried since the attack. All that pent-up energy—the anger, the sadness, the pain—had quietly built up. I had tried so hard on this trip—tried to live it as fully as I could, tried so hard with the challenges I had faced in the past week. Now I was trying not to let it all erupt from within me. I needed to hold on to this one last thing. But I couldn't. My eyes began to blur.

I heard a voice behind me. I was on my knees at the pile, facing away from the kitchen, away from the group, and hadn't heard anyone come up.

"Hey, buddy." It was Mike. He spoke gently.

My emotions plateaued for a moment, holding. My lips still shook, and I felt as if all my insides were quivering. If I opened my mouth, they might just spill out, and it would be all over. It was hard to talk, and I was afraid that if I let anything out, the lump in my chest would burst out in a dry heave. I forced it down just enough to give a short acknowledgment.

"Hey," I said without turning around, trying hard not to break and not to show how close I was to it. If I was going to break down, I wanted to be alone. This was terrible timing. I was about to lose it, and suddenly I had company. I wanted to be somewhere no one could hear me even if I yelled at the top of my lungs. It didn't come out just yet but stayed right there, ready to burst.

There was a beat, and he said softly, "I'm going with ya, buddy."

As my brain processed the words, my upwelling of emotion paused slightly.

"You don't have to do that," I said.

"I know," said Mike, still behind me. I couldn't turn to him.

"Why are you doing that?" I asked him. My throat still quivered, and it wouldn't take much of anything to push me over the edge into a meltdown.

"Because I'm your friend," he said.

With that, my throat stopped quivering, and I let out the breath I'd been holding. The song in my head ended quietly, and the wave I'd been holding back subsided. I breathed. My whole body relaxed, and my eyes sharpened again.

"Thank you," I said over my shoulder but not quite looking at him. With each passing second, I felt my mind and body calming.

"Of course," he said.

After a beat, he turned and left to find Dan. I stayed and finished the last bit of packing, steadying my breath until I was back on an even keel and ready for the next step.

In a few minutes, I had finished, and Dan and Mike came back to the tent. The three of us climbed in to call Camp. Dan had my GPS in one hand and the satellite phone in the other. He dialed Camp, and they picked right up.

"Hey," Dan said into the receiver, "it's Dan ..." He told them Mike and I were ready, and read the coordinates of our campsite from the black-and-white GPS screen. After a minute, he handed the phone to Mike, who said hello and then listened. On a small yellow slip of paper, he wrote down several credit card and phone numbers. I never once spoke on the sat phone. We finished the call and packed it away.

The three of us clambered out of the tent and began packing it. Everyone else was working on cleaning up the kitchen and packing bags. The tent came down quickly, and we stowed it away in no time. Soon, everything was put away. We carefully arranged the pile of Mike's and my gear, including a food barrel filled with trash that we'd bring if there was room.

With all the gear situated, the next thing to do was get ready for the helicopter. Dan gathered flares and smoke and briefed us on how to conduct ourselves around the helicopter. While it landed, we were to stay where we were and not move. Once it landed, under no circumstances were we to approach it from anywhere near the tail. We were to wait for the pilot's instruction and only then approach from the front or the side of the cockpit.

With that out of the way, and the small trove of flares and smoke in a pile next to us, we sat quietly beside the overturned canoes, in the quiet din of wind over tundra. We had set the boats up as a windbreak and waited in the cold wind in our rain gear. A bag of trail mix went around until no one was interested anymore. We talked a bit, but we all were anxious. Dan seemed a bit more relaxed—still uneasy, but no doubt relieved that I was going to get the care I needed and that he wouldn't have to perform any more surgeries.

We talked but would drift off after a sentence or two, once again surrounded by the silence of the wind. None of us knew what to say. The trip would soon be different for Mike and me and different for Dan, Darin, Auggie, and Jean, who had the rest of the trip to paddle. We waited.

I set up for our last group picture, putting the camera on some of our gear and pointing it northwest at our boat, food barrels, and a bit of Thirty Mile Lake. The guys formed a line, all standing but Auggie, who sat on the end of an overturned boat. I focused, turned on the self-timer, and clicked the shutter. I managed to hobble the short distance to the group and stand at the end before the shutter clicked on six tired smiles.

At twelve thirty in the afternoon, a small black spot materialized in the distance to the north. At first, it looked low, coming from the far horizon, but it held a steady altitude. It was a long time before we heard it. It grew steadily, its course never wavering as it drew a straight line toward our solemn group. There was no question where this helicopter was headed: a bearing of 180 degrees, due south from Baker Lake. We had paddled some five hundred miles north and east, and this helicopter would take us even farther north.

Mike and I hugged everyone, and then we all turned to the helicopter.

As it approached camp, Dan raised a small handheld aerial flare and ripped the cord. It shot up with a hissing whoosh, orange sparks drawing an arc into the sky. We were the only colorful things for miles around. Our bright reds and yellows and, of course, our canoes stood out like hunter orange in the sea of drab grasses and steely-blue water. The flare was overkill, but why not?

"I never get to use this stuff," Dan said as he reached for the smoke grenade. It was a small can a third the size of a toilet paper tube, with a little pin on the end. He ripped the pin out and tossed it away from us, in the direction of our landing pad. The truth was, the whole area was flat, with no large obstacles other than our canoes and gear and us. *Everywhere* was a landing pad. The grenade landed with a tiny thud and sat there. No smoke. After a long while, we determined it was a dud, a worthless canister.

As the helicopter approached, we got fidgety. This was it; the group of six was about to be four. Mike and I were about to embark on a separate journey, and so were they. My pulse quickened. This was the first evac for any of the trips I'd been on, and I was the one being extracted. The whine of the helicopter turned to a rhythmic thump as it got closer. When it was close enough that we could just see the pilot, he made a wide arc to his left before following the shore toward us, into the wind.

It was go time.

The helicopter inched in and lowered its altitude. The rotor wash hit us, overpowering the wind at our backs. The pilot eased the controls and kept coming in and down. I could see him through the cockpit's bubble windshield—the typical image of a helo pilot with his mirrored aviators and big mint-green headset. Between us was the pile of our gear, ready to go.

The Bell four-seater's movements were smooth, though the sounds were anything but. We'd been moving under our own power the entire trip, and the loudest sounds we had encountered were the one helicopter landing a month earlier, the far-off hum of a float plane, the roar of the water, the heavy whip of wind, and the terrible sounds of the bear I alone had heard seven days ago. This was louder than all those. The rotors moved impossibly fast, and I half expected them to come flying off from the centrifugal force. They hung on, though, and the pilot inched the craft lower, nearly landing on our pile of gear. *He wouldn't land on the stuff. He's watching, right?* My Pelican case was in the pile, right next to the swaying skid. It had been through so much, and I watched, fearful, as it was about to be crushed by a helicopter skid. I wanted to run up to pull it away. But he touched down on the grass two feet from the case.

We walked in a wide arc around the helicopter until we were off his starboard side, and waited there. Once we'd gotten there, the dark-blue cockpit door swung open and the pilot set a boot down on one of the skids. He looked comfortable under the rotor wash, aware of the thrum a few feet above his head but unworried by it. He had taken off the headset but still wore the aviators, walking with lowered head and shoulders, modestly shielding himself from the wash. He wore a new-looking Carhartt jacket and heavy pants. His sunglasses looked immaculate, and he hid his hands in the pockets of his coat. As he came closer, I noticed a sheath on his belt. It was a Leatherman Wave, same as mine. Everything about him looked new, spotless.

When he was close enough, he yelled over the noise of the helicopter, "One of you guys got attacked by a bear?"

"That would be me," I yelled, raising my hand and taking a small step forward from our line. He eyed me from head to toe, shrugged, and motioned us to the chopper. Mike, Dan, and I stepped forward after we waved to the guys.

The pilot instructed Mike and me on how to get into the helicopter and how to put on our seat belts and headsets once we were inside. He turned to Dan. "I'm supposed to give you this battery, too," he said, pulling from his jacket a brick the color of our satellite phone.

Dan produced a similar one from his pocket and handed it to him. "Here's my empty one."

I said goodbye to Dan and stepped around the nose of the helicopter. Grabbing my Pelican case from near the skid, I opened the door and climbed awkwardly into the front seat next to the pilot. By the time I had my seat belt on, Mike was in back with the portage pack, the food barrel, and our paddles, ready to go. As soon as I put on the headset, the din of the helicopter quieted tremendously, as if I'd dipped underwater. The pilot checked Mike's harness and muffs and came around to do the same with me before he took his own seat at the controls. I pulled out my camera and stowed the case for the flight.

Dan, Auggie, Darin, and Jean had grouped together in front of the helicopter. They stood near the canoes and the food barrels, motionless, watching through the Lexan windshield as we situated ourselves, waiting in the rotor wash for our takeoff. They suddenly looked very alone. From the cockpit of the helicopter, the tundra looked much emptier, and we weren't even off the ground yet.

The headset buzzed, and the pilot's voice scratched through. He welcomed us aboard, gave us a quick briefing, and told us that our headsets had microphones, so we could talk with him. We acknowledged this, and he began his preflight checks. After a moment, the pilot pushed on the throttle, and the main rotor revved even faster.

Outside the helicopter, the guys sensed the change. They raised their hands in a somber wave. I waved back, and we lifted off. We were suddenly high above the ground, looking down through clear windows at our canoes, our canoe mates drifting off into the flat of the tundra. They waved until they became small dots of color, until we turned completely around and could no longer see them.

Below us, the grasses and tussocks of the tundra disappeared, blending into an olive-drab carpet. The rotors thumped, and the helicopter shook, and we made our heading zero degrees: due north. The hills smoothed to gentle rises, and Thirty Mile Lake shrank into a long, dark form. Suddenly, we could see all around us, hundreds of miles in any direction. We were above an endless expanse of lakes, waterways, and tundra that stretched beyond the horizon.

The pilot turned to me, and I heard the headset sizzle through the whine of the helicopter. His voice was muffled, artificial-sounding. "So what happened?"

CHAPTER TWENTY-THREE

Premonition

I was ten years old, in bare feet and swim trunks, walking deeper and deeper into the cold, dark forest, along a narrow path up and down ravines. Every breeze sent chills across my skin. Autumn was approaching. The night had been calm, with chilled air that transmitted every sound with astounding clarity and carried with it a tension that foreshadowed the coming storm.

I followed the path and the pair of feet in front of me until we crested the last rise and saw it. In the sheltered hollow of a ravine was a low building, primitive but stout. The small dome had a round door covered with skins. Before the door was a great hearth of round stones. This was the sweat lodge.

Before long, the fire was lit. The night had deepened, and a dry blowing storm shook the forest around us. Our small ravine was dark but for the white bursts of lightning, our crackling fire, and the glowing stacks of large round stones at its edge. Silhouetted against the orange glow were the rest of our sweat ceremony: students and an elder from the Ojibwe tribe—a holy man. Behind them, the sweat lodge sat low, its small opening protruding slightly like the door to an igloo. My family and I were guests here.

The fire crackled as we passed the glowing rocks and crawled into the sweat lodge. Inside the only light came from a pile of orange rocks radiating heat from a depression in the center. I watched the deep glow of the hot stones and imagined they were dragon's eggs. We perspired for a long time to the sound of chanting and drumbeats.

When the holy man pounded the last beat of his drum, one of the stones cracked under the intense heat. A spray of sparks shot into the air, illuminating the small, dark space for an instant. There were not many of us in the lodge, and no one sat across from me, yet the sparks formed the distinct outline of a figure, seated cross-legged opposite me and holding a pipe. Just as suddenly as the rock had cracked, the sparks disappeared, the constellation was gone, and the lodge was again dark. I sat back, startled, and looked to see if anyone else had noticed. Everyone sat still, silent. The image was burned in my vision, like the flash of a camera, until the silence ended, the skin was pulled back from the entrance, and we emerged into the calm night.

Outside the lodge, the storm was gone and a blanket of stars hung over the night sky. I hesitated for a long time before finally approaching the elder. I told him what I had seen in the lodge. He listened quietly, nodding slowly. The deep wrinkles of his face were like tree bark in the firelight. After a long, thoughtful pause, the old man spoke.

"You have a spirit companion," he said, his voice quiet and deep like distant thunder. "Mukwa," he said. "Do you know what 'mukwa' means?"

"No," I said.

"'Mukwa,'" he said, eyes fixed on mine, "is Ojibwe for 'bear.'"

EPILOGUE

The afternoon Mike and I flew north in the helicopter, Dan and the guys caught up with the Femmes. A few days later, both groups paddled into Baker Lake.

By that time, we were long gone.

When we arrived at Baker Lake, we had learned that there was no rabies serum, nor enough antibiotics to completely stop the infection. So after twelve hours of IV antibiotics, EKGs, and interviews with the RCMP and the game warden, Mike and I were shipped south to Winnipeg. There, I continued the antibiotics and began my course of postexposure rabies shots. My family drove through the night to pick us up, and we made our way back across the border to Menogyn.

With my leg showing steady improvement over the next several days,

Mike and I were cleared to go north with the pickup crew to meet the Hommes and Femmes in Winnipeg. After hugs at the airport, we came back to Menogyn to a grand ceremony and the welcoming arms of our families. We enjoyed a feast, shared stories from our adventures, and hung our red Hommes paddle next to the others on the wall of the dining hall.

At the end of each Menogyn trip, we exchange bracelets tied from parachute cord, to remind us of each trip we've taken and to recognize camper and guide alike for their accomplishments and growth. Looking for a private spot to cap off our trip, we made our way to the boathouse dock and the twenty-five-foot voyageur-style boat used to ferry campers back and forth across the lake. Its bow and stern rose sharply from the water, in crescents like the moon that had yet to rise over the calm lake.

We untied the boat and pushed off from the dock, paddling east out of the bay and into the still waters of West Bearskin Lake. Soon, we were far from camp, paddling through the reflections of millions of stars. We stopped and let the boat float, nearly motionless. Each of us lay down across the bottom of the boat. I marveled at the blanket of sparkling starlight.

Normally, we each would take a cord, pass it to someone we wanted to acknowledge and express our appreciation for, and we would share with the group the good qualities we had noticed from that person, how they had changed, how they had overcome obstacles, and how they had grown or helped us grow. The recipient would take the cord with a modest smile and wear it on their wrist to remember the trip. That night, though, the six of us lay looking up at the stars as they turned slowly around our gently drifting boat and around our turning globe. We weren't silent; we each spoke in turn, quietly to preserve the quiet spell of the lake. Mostly, though, we lay listening to the lap of water and the soundless spin of stars. What more was there to say?

Two weeks after leaving Menogyn, I started my senior year of high school. Two months after the bear attack, on the last day of September, my wound finally closed. Every day, my injured leg grew stronger, yet months later, it could do only 75 percent of what the left leg could do.

The next summer, a full year after the attack, I went to physical therapy, doing exercises and going through painful massage to break up scar tissue. Eventually, I regained my range of motion. After that, it was just a matter of time before my strength came back, too.

The muscle would never regain its full capacity, nor would the feeling on the front of my thigh return. When I tighten my hip flexors, the large scar slides down my leg with the flexion of muscle, pulling the skin into a divot half an inch below the surface. Around the scar and radiating down my thigh is a spot the size of a softball where the tissue is numb. Sensations come from it as dull prods, if at all.

Early on, I decided not to let my experience stop me from doing what I want to do. It hurts nearly every day, but as with other things in life, it has become an everyday expectation, part of my experience. It reminds me that I am lucky to be here. Years afterward, I went back to Menogyn to work as support staff in camp and later as a trail guide, leading kids on transformational wilderness experiences like the ones that had formed me.

After the attack, I graduated from high school and college and moved to northern Minnesota, where I now volunteer with search and rescue in the woods and wilderness I grew to love. I accidentally brought up the bear story on a first date, and she didn't believe a word of it. Four years later, we were married at Spirit Mountain, overlooking Duluth, the St. Louis River, and Lake Superior.

I think often about what happened, about the bear and the tundra and the people who helped me. Sometimes, I remember those fleeting moments, too, those little bits of memory that lie hidden. Like my conversation with the game warden in Baker Lake. "We'll be going out to Princess Mary Lake," he'd said. "We're going to try to find the bear and see if it reacts hostilely to humans."

"How are you going to do that?" I asked.

"We'll take a helicopter out there, find the bear, drop my friend here on the ground," he said, nodding to the man sitting beside him. "And if it reacts negatively to him ..." he paused, "... we'll destroy it."

"It seems dangerous to put him on the ground with the bear," I said.

"He's done this before," the warden said, looking to his companion. His partner turned to me and spoke. He told me the story of a problem bear they'd had not far from Baker Lake. He had gone out to find it, and it charged at him from a great distance. He raised his rifle and fired a shot. The bear had been in midgrowl. The bullet entered through the roof of the open mouth, exiting through the bear's skull and dropping it to the ground in midstride.

"We'll be flying out to Princess Mary tomorrow," the warden said. "We'll see what happens."

But the bear was never found.

It's better that way. I like to think that it disappeared that morning, with the fog on Princess Mary Lake.

ACKNOWLEDGMENTS

There are so many people who have had a hand in this story, and in writing an acknowledgments page, I risk the possibility of forgetting someone, which gives me pause—this book simply wouldn't be, if it weren't for those that helped me along the way.

I'd like to thank my family, and my parents who always believed I could write this book, who had the courage to put their trust in others as all of this was unfolding, and for their help with edits, ideas, and feedback throughout the process. I'd like to thank Dan, whose expertise and care kept me safe and helped to get me through, and for his perspective afterward, which helped to turn this story into what it is. I'd like to thank Mike for his friendship, then and now, and for cutting his trip of a lifetime short for me. I'd like to thank Paul and Aaron, the health officer, and everyone involved with Menogyn. It truly is one of the best organizations I've ever had the pleasure of working with. I appreciate what they've done for me, how they handled this situation, and the growth I've experienced there as a camper and staff. I'd like to thank the other guys on our trip, Auggie, Jean, and Darin, for their friendship on that trip. I want to thank the helicopter pilot, whose name and association I don't know. I want to thank the people of Baker Lake who I met and worked with, including the staff of the Baker Lake Health Centre, the folks from the Nunamiut Lodge Hotel that hosted Mike and me, the officers from the RCMP, the game warden and his staff, and the people at Baker Lake Airport and Calm Air International airlines. I'd like to thank the staff at Health Sciences Centre Winnipeg, North Shore Health Hospital in Grand Marais, and all of the experts with whom Camp consulted, including those from the CDC and specialists in infectious disease.

I am so appreciative of the help from my agent, Philip Turner, who saw potential in my story and text early on, and worked with me to edit this book at those beginning stages when the text was still raw. I'd like to thank my editor, Michael Carr, for the astute work of refining the book to what it is today. I'd like to thank Blackstone for believing in this book. I'd like to thank some of my earliest readers for their insight and support: Catherine Watson, Danny Naslund, Amy Freeman, Dave Freeman, and Jim Brandenburg.

Finally, I'd like to thank my wife, Lacey. Without her steadfast help, never-ending reads, rereads, edits, and patience, as well as persistent encouragement, this book simply wouldn't be.

GLOSSARY

Arctic: Relating to the North Pole and all regions typically above the Arctic Circle (currently at 66°33'47" north of the equator).

backward ferry (or back ferry): A white-water technique to move laterally across flowing water, wherein the canoe is back-paddled with its stern upstream and at an angle toward the desired bank, allowing the flowing water to help push the craft perpendicular to the flow.

back paddle: A technique used to move the canoe (or other watercraft) backward, typically by inserting the paddle into the water behind the paddler, then pushing it forward.

bow: The front of the boat.

cross draw: A draw performed on the nondominant paddling side, in which the paddler reaches over to the side she or he was not paddling on, without switching hands, and draws on the opposite side.

downstream V: A white-water feature between two other white-water features, where the water flows quickly and generally smoothly. Downstream V's are most often the most easily navigated feature in white water and are sought when choosing a route.

draw: A paddle stroke in which the paddler "pulls" water toward the boat, moving the boat toward the side on which the stroke is being performed.

dufek: A rudder technique used only in the bow of the canoe, wherein the bowman places the paddle at the very front of the boat. This is an aggressive technique and can cause the boat to turn very quickly, and in some cases can flip the boat if done improperly or if both paddlers are not ready for the sudden shift.

eddy: A circular movement of water, counter to a main current, causing a small whirlpool and typically found on the downstream side of a rock or other feature.

esker: A long ridge of gravel, sediment, and scree, typically following a winding course, deposited by meltwater from a glacier or ice sheet.

feature: A distinct obstacle in white water, or distinct attribute on land.

ferry: A white-water technique to provide lateral movement across flowing water.

forward ferry: A white-water technique to provide lateral movement across flowing water, wherein the canoe is paddled forward with its bow pointed upstream toward the desired bank, at an angle to the current, allowing the flowing water to help push the craft perpendicular to the flow.

freeboard: The distance from the water's surface to the upper edge of a boat.

GPS: Global positioning system, an accurate worldwide navigational and surveying facility based on the reception of signals from an array of orbiting satellites. Often used as shorthand for a handheld navigational-aid device that uses the system to provide accurate location measurements.

matt food: A protein-rich food consisting of peanut butter, oats, powdered milk, and honey.

outback oven: Cookware for baking on a stove, comprising a wide skillet with a tight-fitting lid, a thermostat, a heat-capturing fabric tent, and heat-dispersion devices for the bottom.

paddle: A tool made from plastic, wood, fiberglass, or other material, used to move a canoe or other boat through water. It differs from an oar in that it is not attached to the watercraft.

pemmican: Historically, a paste of dried and pounded meat mixed with melted fat and other ingredients, originally made by Native North Americans and later adopted by Arctic explorers. When referred to in this text, it is a dry mix of seeds, oats, and other nutrient-rich foods, similar to trail mix.

pillow: A white-water feature in which water is pushed over an obstacle and forms a bulge where the water does not break until downstream from the obstacle. These are difficult to see from upstream.

port: The side of a ship or aircraft that is on the left when one is facing forward. The opposite of starboard.

portage: The carrying of a boat and its cargo between two navigable waters.

Prophet: A type of durable three- or four-person dome-shaped backpacking tent made by Mountain Safety Research, with an igloo-style crawl-space door in the rainfly. Designed for all four seasons and weighs thirteen pounds.

pry: A paddle stroke that pushes the canoe away from the side on which the stroke is being performed. Typically done by inserting the paddle and pushing, or prying, water away from the boat.

recycler: A white-water feature where waves form and repeatedly or continuously crash in a tumbling motion. Recyclers can trap objects and tumble them for long periods and, when big enough, can do the same with people and watercraft.

rod: The linear unit of measure used to describe portage length, equal to 16.5 feet. For reference, 320 rods equal roughly one mile.

rudder: A static paddle movement that uses the paddle as an angled plane against the momentum of the boat or current to adjust the course of the canoe.

sideslip: A paddling technique that moves the canoe laterally without the aid of the current, typically through repeated draws or pries.

starboard: The side of a boat, ship, or aircraft that is on the right when one is facing forward. The opposite of port.

stern: The back of a boat.

strainer: A feature in white water, typically a tree or root system but occasionally a natural placement of rocks, that allows the current to pass through. Strainers are dangerous for paddlers because they admit enough fast-moving water to trap objects, including people and watercraft.

taiga: The often-swampy coniferous forest of high northern latitudes.

tundra: The vast flat, treeless Arctic region of North America in which the subsoil is permanently frozen.

white water: A fast, shallow stretch of river where the rocks, shore, and river bottom directly affect the surface, forming waves and other obstacles.

AUTHOR'S NOTE

This is a memoir, a work of nonfiction, written with every effort toward accuracy from firsthand memories, dutiful journaling, photographs, and interviews with key parties afterward. When the perspective shifts from that of the narrator, the details included have been extracted from interviews with those who were there.

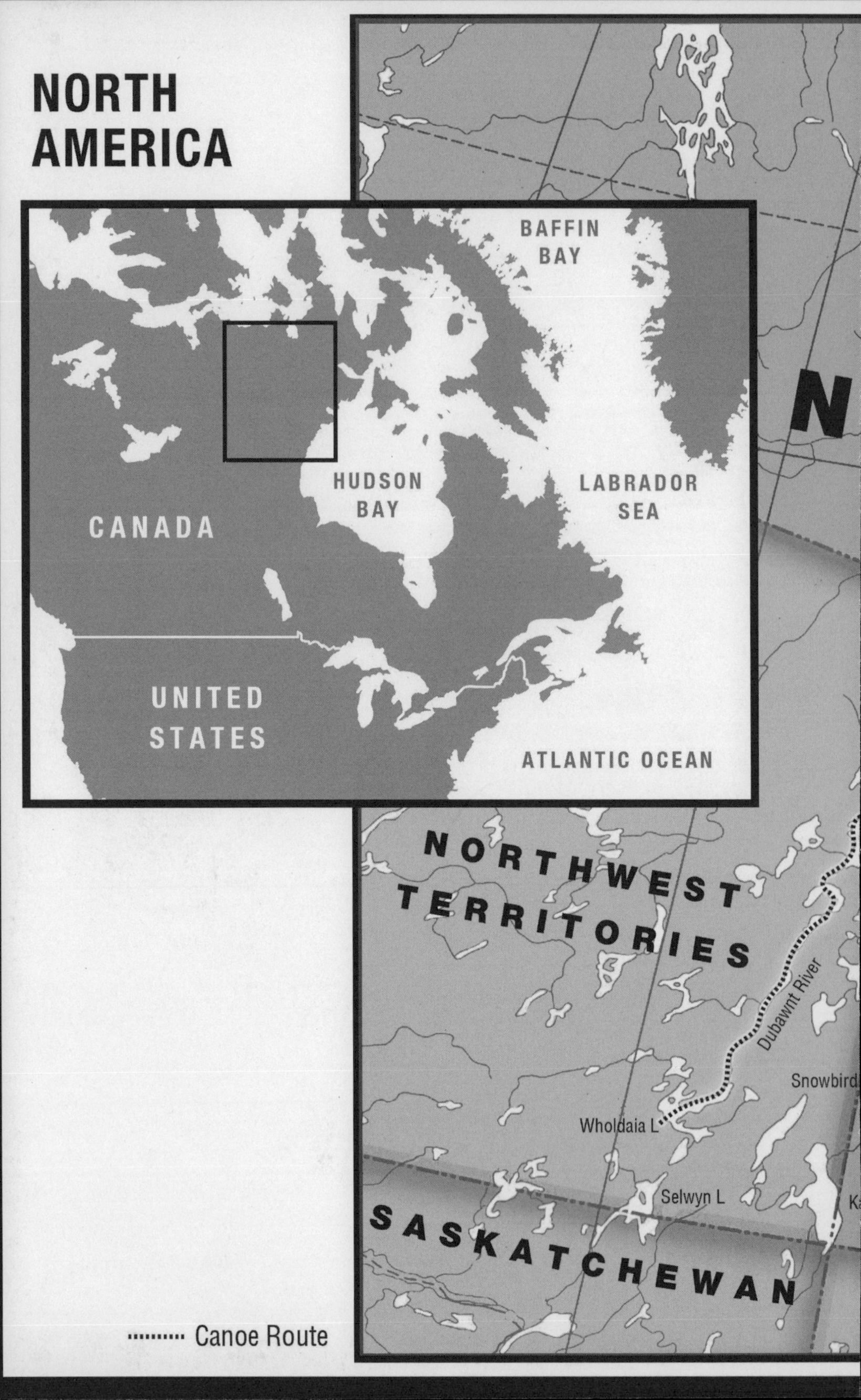
NORTH
AMERICA
BAFFIN
BAY
HUDSON
BAY
LABRADOR
SEA
CANADA
UNITED
STATES
ATLANTIC OCEAN
NORTHWEST
TERRITORIES
Dubawnt River
Snowbird
Wholdaia L
Selwyn L
SASKATCHEWAN
Canoe Route